# A Study in Genesis

## From Adam to Abraham

## M. J. Tiry

M J Tiry

ISBN: 979-8-9903305-0-4-.Paperback
ISBN: 979-8-9903305-2-8.Epub
ISBN: 979-8-9903305-1-1 Casebound

Library of Congress Control Number 202-490-6933

Print Information Available on the Last Page

Published by: M J Tiry Publishing 2024

i

# PREFACE

## Supernatural Creation

The concept of supernatural creation is dear to the heart of every Bible believer. However, in this writer's personal experience, a person will not be interested in making a serious study of the concept of a supernatural creation until he or she comes to the point in life of dealing with his or her eternal destiny. It is the hearing of the concept of the God of creation entering into our human race to go the cross of Calvary to pay sin's debt in full so as to offer to anyone and everyone the gift of eternal life that changes things. It is that conviction that changes one's perspective on the subject. It is then that a person's interest in the veracity of the Word of God is piqued by the prospect of having eternal life as a gift of God's grace. It is then that he becomes genuinely interested in the origins of life.

The study of Genesis is an important part of the believer's view of the world and then having an understanding of the eternal purposes of God and God's plan for the ages. As we study such subjects as the Kingdom that God promised to Israel in the Old Testament or the Church which is Christ's Body in the Pauline epistles, we need a framework for putting these concepts into perspective. The Book of Genesis gives us that framework. Further, it gives us understanding of why the world is the way it is and why God has taken the action that He has in the redemption of man. It also sets out for us the basic conflict between God and Satan that has raged through the ages.

We ask: "Can creation be proven by science?" The simple answer is "No!" There are four requirements for the scientific method of verification to be used to verify a theory by empirical science:
1. Observation – You have to be able to collect empirical data.
2. Experimentation – You have to be able to study the theory in a laboratory.
3. Reproduction – You have to be able to reproduce the results.
4. Falsification – The theory must be able to be tested and proven to be false.

None of these requirements are met in the examination of the empirical data regarding creation.
1. Neither creation nor evolution has ever been observed by any human being (Adam excepted).
2. Since the phenomenon of the created universe far exceeds the lifetime of any experimenter, laboratory study of it is not possible.
3. Neither creation nor evolution can be reproduced in the laboratory. Therefore, the study of it is not in the realm of experimental science.
4. Neither creation nor evolution can be refuted by science, thus it is outside the realm of empirical science but are in the realm of faith.

However, this does not mean that the believer is without a scientific basis for his belief in supernatural creation over other theories. To demonstrate this, let's summarize all of the theories that have ever been put forth on the existence of the matter and energy of the universe. There are four basic theories that people put forth on the origin of the universe:

1. The universe has a mind of its own and created itself.
2. It has always existed (i.e. that matter and energy existed from eternity past.)
3. It does not really exist.
4. It had a supernatural creation by an omnipotent, eternal, omniscient, omnipresent God who created it out of nothing. (The presentation that we have in the Bible.)

Now let's examine these theories in light of the laws of science that govern this present universe.

1. The first law of thermodynamics (for simplicity think of it as the first law of science) states: "Matter and energy can neither be created nor destroyed." We understand from nuclear physics that matter and energy are interchangeable (i.e.. we can convert mass to energy in nuclear reactions according to the Einstein equation of $E=mc^2$) but the sum total of matter and energy remains constant. This tells us that according to the laws that govern this present universe, that the universe could not have created itself. Thus we eliminate theory #1 above.

2. The second law of thermodynamics states basically: "In every energy transformation process, a certain amount of energy goes into its waste form of heat." This law of science implies that if the universe were to go on into infinity – in to eternity (i.e. if it were to go on forever), then all energy will eventually go into its waste form of heat. In order for motion to occur, energy has to be at two different levels. If all energy were in its waste form of heat, then all energy would be at the same level and no further motion can occur. This eliminates the theory that the universe has always existed. If that was true, then all of the energy should have gone into its waste form of heat and all motion should have stopped. In scientific terms, the universe would have died a "Heat Death." The fact that motion is still possible in this universe, we understand that the universe that we see and observe has not always existed.

3. The third possibility listed above (i.e. that it does not actually exist) is the basic definition of insanity – to be not in touch with reality. Therefore, we dismiss this out of hand.

4. The fourth possibility listed (that the universe had a supernatural creation by an omnipotent Creator) remains the only possibility that is consistent with the observable phenomena of science. Further, as we study the Book of Genesis, we find that it is consistent with what we observe in the natural sciences. That is: the Genesis account provides a completely satisfactory explanation for all that the believer observes in the real world.

- In the account of the flood we have a satisfactory explanation for the existence of the sedimentary geology: limestone, sandstone, and shale deposits and the fossil record. We will study this subject further when we come to chapters 6, 7 and 8 of Genesis.

- As we consider the division of the earth in the days of Peleg and compare it with scientific information in the Book of Job, we find a satisfactory explanation for the glacial geology that we see around us here in the northern latitudes. We will study this subject further when we study Chapters 10 and 11 of Genesis

- As we study the events associated with the tower of Babel, we find a satisfactory explanation for the origin of the various languages.

- As we study the family of nations in Genesis 10, we have an explanation for the origin of nations as we see them today.

- As we study God's interest and activities in separation of one nation (Israel) from all of the other nations and observe His dealings with that nation, we understand why the world is as it is today. And we can understand the prophetic future of that nation specifically and humanity in general.

- As we study the creation of man in the image of God, the nature of man as a free moral agent, the failure of man, and the work of God as He acts in a benevolent manner to redeem fallen man, we understand that God has a purpose for man that He reveals in the Scripture. That purpose is twofold:
  1. He has a purpose for the earth that centers in Israel. That purpose involves His Son reigning on the earth (Jer. 23:5) through the redeemed nation of Israel (Rev. 5:10).
  2. He has a purpose for the heavens that centers in the church which is Christ's body and is the subject of the mystery that God hid in Himself (Eph. 3:9) until the time was right to reveal it (1 Tim. 2:6). When He revealed this mystery, He did it through the apostle Paul (Eph. 3:3) – the apostle of the Gentiles (Rom. 11:13).

The simple statement, "In the beginning God created the heaven and the earth" refutes:

- Atheism – because God created the heaven and the earth by bringing both into existence when nothing existed before either.
- Pantheism – because God is separate from His creation. The universe could no more create itself than the computer that I am working on could have created itself.
- Polytheism – because One God created all things.
- Materialism – because all matter had a beginning.
- Dualism – because the Triune God was alone when He created all things that have been created.
- Humanism – because God, not man, is sovereign and is the ultimate authority.
- Evolution – because God created everything in the space-matter-time continuum of the universe out of nothing and fixed it so that the organic creation could reproduce only after its kind.

## Notes to the Student

### Subject Outline

Before entering into an in depth study of any book of the Bible, one should first read the book a number of times and then prepare an overall outline of the book. A subject outline of the book of Genesis is helpful to enable you to see the flow on thought and to provide a framework for your study of the book. We encourage you to first prepare your own outline of the book of Genesis after several readings of the book.

### Use of a Structural Analysis

This study in its original form provided a structural analysis of each of the first twelve chapters of the book of Genesis. A structural analysis is made by laying out the main point of a passage across the page and then placing the subordinate clauses either above the line of the main point or below it depending on whether it comes before or after the main point in the sentence. A series of three dots (…) is then placed in the text where ever words were moved. Finally, a line is drawn to connect the phrase to the point of its original location. By making use of a structural analysis of the text, you will be able to quickly identify and understand the main point of the verse or passage. Once the main point is understood, the subordinate points will add important details to that main point. The point or thought of a passage has two parts: the subject and the complement. The subject can be identified by answering the question "What is the passage talking about?" The complement can by identified by answering the question "What is being said about what is being talked about?"

The idea or thought of the passage can be discovered by asking every question that can be raised about the passage and then answering these questions from the text, the context, and the other parallel passages that address the same subject. To come up with the right questions, apply the six interrogatives (What? Where? When? Who? How? and Why?) to the passage under consideration. Ask yourself every question that you can think of regarding the passage and write those questions down on a sheet of paper. Then write your answers to those questions. . The benefit of this exercise is that you can focus on the main point that the Holy Spirit is making in the passage under study. The annotations on Chapter 10 of this study presents and example of a structural analysis of the Bible text.

### Regarding Use of Cross References

A good cross reference will be helpful in the course. We recommend the *Treasury of Scripture Knowledge*. The *Treasury of Scripture Knowledge* is simply a compendium of cross references for each verse of the Bible. The Bible is the best commentary on itself that you will find. By tracing the thought through the cross references, you will find that the Bible teaches itself. The cross references will take you to parallel passages that address the same subject as the passage in question.

### Annotations

In studying a passage, you ask the interrogative questions as shown above. After formulating your questions, begin to answer them by a study of the text in its context as well as searching out the parallel passages and cross references. Also, use the cross reference (e.g. *Treasury of Scripture Knowledge*) to compare scripture with scripture. As you answer your questions, write out your answers and reference the verse that the question came from. For the benefit of you the reader, this author's annotations are provided to you.

### Appendix

Some of the various annotations will refer you to the appendix. The appendix is contained at the end of this study.

### Study Guide Questions:

Study Guide Questions are provided at the end of each chapter. To get the most out of this study, it is recommended that the reader review the study guides questions before reading the chapter. Answers to those questions can be found in the chapter in sequential order. It is recommended that the Bible study principle of Act 17:11 (searching the scripture to see if these things are so) should always be employed. On a separate paper, make notes of what should be checked out for further personal study.

### Appreciation

I have enjoyed the study involved in putting this material together. I do appreciate the diligent work that others before me have put into the study of Genesis. I hope that each reader and student who picks this study up enjoys it as much as I did in preparing it. I certainly do not claim to have the final word on this important book of the Bible but I offer what I have learned in the hopes that it is edifying to others. I just ask that you who go through this study will give it the Berean test of searching the Scriptures "…to see if these things are so" and to then get back to me with your feedback.

I thank my daughters Naomi Salgado for her work in typing this material and in preparing it for publication, Leah Lietz for her work in formatting and Anna Anderson for her work with the graphics. I thank my wife Linda for her patience while I set the list aside for the effort to put this together.

A word is in order here on how to study the Bible and actually how to approach the Bible. Some basic principles to hold in our study of Scripture are:

1. All of scripture came from the mouth of God and it fully equips the man of God to do anything that God would have him do. "All scripture *is* given by inspiration of God, and *is* profitable for doctrine, for reproof, for correction, for instruction in righteousness: That the man of God may be perfect, throughly furnished unto all good works." (2Timothy 3:16-17) It can be said that the Holy Spirit never works apart from the Bible and the Bible never works apart from the Holy Spirit.

2. The term "inspiration of God" in 2Timothy 3:16 means that it was breathed out of God's mouth. It is truly as the Lord tells the devil in Matthew 4:4 "It is written; Man shall not live by bread alone, but by every word that proceedeth out of the mouth of God." And that is the origin of every word of scripture – from the mouth of God.

3. Scripture must be studied in its context in order for it to make sense. There are two contexts: the immediate context in which the passage is set and there is the remote context that looks at the Bible as a whole. Billy Sunday is said to have made the statement "A text without a context is a pretext." That concept is what Peter is communicating when he said in 2Peter 1:20 & 21 "Knowing this first, that no prophecy of the scripture is of any private interpretation. For the prophecy came not in old time by the will of man: but holy men of God spake *as they were* moved by the Holy Ghost." No passage of scripture is intended to stand by itself but rather each passage actually relates to every other passage of scripture. One of the greatest tools to Bible study is a good cross reference (As noted above one that I like is *The Treasury of Scripture Knowledge*). By comparing scripture with scripture the Bible teaches itself. The Bible itself is its greatest and best teacher.

4. While all of scripture is written for our learning, not every passage of scripture is specifically addressed to us. The word of truth then must be rightly divided. Paul tells us this in 2 Timothy 2:15 saying "Study to shew thyself approved unto God, a workman that needeth not to be ashamed, rightly dividing the word of truth." We trust that appendix 4 of this book will be very helpful in seeing this concept.

5. Another key to understanding the Bible is simply to let it say what it clearly says. It is a major mistake to spiritualize scripture. The Bible is written to be taken literally. There are times when the Bible uses figures of speech (figurative language) but when it does it is apparent that such is the case. Basically, we must remember the adage "if the literal sense makes perfect sense, seek no other sense."

6. God has taken great care to give us His inspired word and gave it without error. He has also pledged to preserve it so. (Psalm 12:6-7) "The words of the LORD *are* pure words: *as* silver tried in a furnace of earth, purified seven times. Thou shalt keep them, O LORD, thou shalt preserve them from this generation for ever." It is the conviction of this author that there exists today a preserved text of the inspired and inerrant Word of God. This is not in the original auto graphs for they have been lost

through time but this preservation of scripture is in the multiplicity of copies. It was God's desire and design that the Bible gets into the hands of the people. If there is a doctrine of preservation, then that preservation is done in the multiplicity of copies. This author holds the conviction that the preserved text line is the Received Text (Majority Text) of the New Testament and the Masoretic Text of the Hebrew Old Testament. Since there is only one translation in print in English today from these, all scripture references in this study are from the King James Bible.

7. There is yet another key to an effective study of the word of God. That is the heart attitude of the Bereans of Acts 17:11. They received the Word with and open mind but they did not just take any man's word for truth or error of what was said until they searched it out in the scripture. That approach gave them protection from error for they made the Word of God their final authority and examined what everyone said based on the Word of Truth – the Bible.

8. One final thing regarding the Bible having the impact in our lives that God intended it to have is the simple matter of believing it. Paul tells the Thessalonians that they received the word of God, they received it not as the word of men but as it is in truth the Word of God, which "…effectually worketh in you that believe." It is not just the understanding of it that makes it effective but applying it by faith to one's life that makes it effective to give spiritual strength and vitality.

M. J. Tiry

# The LORD by Wisdom hath Founded the Earth:

## Proverbs 3:19-20 (KJV)

<sup>19</sup> The LORD by wisdom hath founded the earth;
by understanding hath he established the heavens.
<sup>20</sup> By his knowledge the depths are broken up,
and the clouds drop down the dew.

# Wisdom Personified in Creation

## Proverbs 8:12-36 (KJV)

<sup>12</sup> I wisdom dwell with prudence, and find out knowledge of witty inventions.
<sup>13</sup> The fear of the LORD *is* to hate evil: pride, and arrogancy,
and the evil way, and the froward mouth, do I hate.
<sup>14</sup> Counsel *is* mine, and sound wisdom: I *am* understanding; I have strength.
<sup>15</sup> By me kings reign, and princes decree justice.
<sup>16</sup> By me princes rule, and nobles, *even* all the judges of the earth.
<sup>17</sup> I love them that love me; and those that seek me early shall find me.
<sup>18</sup> Riches and honour *are* with me; *yea*, durable riches and righteousness.
<sup>19</sup> My fruit *is* better than gold, yea, than fine gold; and my revenue than choice silver.
<sup>20</sup> I lead in the way of righteousness, in the midst of the paths of judgment:
<sup>21</sup> That I may cause those that love me to inherit substance; and I will fill their treasures.
<sup>22</sup> The LORD possessed me in the beginning of his way, before his works of old.
<sup>23</sup> I was set up from everlasting, from the beginning, or ever the earth was.
<sup>24</sup> When *there were* no depths, I was brought forth;
when *there were* no fountains abounding with water.
<sup>25</sup> Before the mountains were settled, before the hills was I brought forth:
<sup>26</sup> While as yet he had not made the earth, nor the fields,
nor the highest part of the dust of the world.
<sup>27</sup> When he prepared the heavens, I *was* there: when he set a compass upon the face of the depth:
<sup>28</sup> When he established the clouds above: when he strengthened the fountains of the deep:
<sup>29</sup> When he gave to the sea his decree, that the waters should not pass his commandment:
when he appointed the foundations of the earth:
<sup>30</sup> Then I was by him, *as* one brought up *with him*:
and I was daily *his* delight, rejoicing always before him;
<sup>31</sup> Rejoicing in the habitable part of his earth; and my delights *were* with the sons of men.
<sup>32</sup> Now therefore hearken unto me, O ye children: for blessed *are they that* keep my ways.
<sup>33</sup> Hear instruction, and be wise, and refuse it not.
<sup>34</sup> Blessed *is* the man that heareth me, watching daily at my gates,
waiting at the posts of my doors.
<sup>35</sup> For whoso findeth me findeth life, and shall obtain favour of the LORD.
<sup>36</sup> But he that sinneth against me wrongeth his own soul: all they that hate me love death.

# Contents

PREFACE
INTRODUCTION


ANNOTATIONS

# Figures

# Appendices

TABLES

# Abbreviations used:

| | |
|---|---|
| c. | circa ("about/Approximately") |
| cp | Compare |
| e.g. | exampligratia ("for example") |
| et. al. | Et allii ("and others") |
| etc. | Et. cetara ("and so forth") |
| ff | and the following (verses, pages, etc.) |
| i.e. | id. est. (that is) |
| vas, vv | verse (s) |
| viz. | Videlicet ('namely") |

# References cited:

Brook, Robert. Date unknown; Illustration "Abraham the Father of us All"
Jones, Floyd Nolen, 21st Edition, 2019 *The Chronology of the Old Testament*, Master Books
Patton. Donald W., 1966, *The Biblical Flood and the Ice Epoch*, Pacific Meridian Publishing Company
*General Biblical Introduction From God to Us*, H. S. Miller (1960)
Tiry, Michael J, 2021, *You and Your Creator*, WordHouse BP
Tiry, Michael J, 2023, *A Study in the Revelation*, M J Tiry Publishing
Capital Bible Seminary, *The Treasury of Scripture knowledge*, 1982, MacDonald Publishing Compny, Inc.

# INTRODUCTION

## A Study in Genesis – From Adam to Abraham

This study in Genesis covers the first twelve chapter of Genesis. It is a verse by verse expository presentation of the text of Genesis with annotations on each verse. The verse by verse expository study of the book actually starts on page 25. In this preface, we will look more in depth at the first verse of Genesis because it serves as a prelude to the events that are outlined in the rest of chapter one.

Genesis is the account of the creation of man in the earth. The focus of Genesis as the record of the origin of man in the earth has a back story that is contained in the later chapters of the Bible. We will look at that back story here in the preface and then proceed to study the events of what we will call creation week – a literal six day period of six days of 24 hours duration each in which the world that we live in came into being.

## The Creation of Heaven and Earth

> ¹ In the beginning God created the heaven and the earth. ² And the earth was without form, and void; and darkness *was* upon the face of the deep. And the Spirit of God moved upon the face of the waters.
>
> (Genesis 1:1-2)

## Genesis 1:1 & 2

Verse 1 tells us what God created in the beginning. He created the heaven (note that this is heaven in the singular) and the earth. The heaven would be the stellar universe and the earth would be the earth in the form it was in at the beginning of time. Then in verse 2 we find that the earth was without form and void and that it was under water that is called "the deep." We ask at this point "Was it without form and void because God had not taken it through creation week yet or was it in that state because it became that way?" That is a very relevant question because the word translated "was" in verse 2 is also translated "become" and "became" in Genesis 2:7; 18:18; and 19:26. Also, in Genesis 1:26 and 27, we find that when God created man in the earth He commissioned man to "replenish the earth." This replenishing of the earth implies that the earth was once plenished and had then subsequently been depopulated. This compels us to ask "Who was it that once populated the earth and then left it or was removed from it?" Add to that, the term "without form and void" is from the Hebrew words "Tohu" and "Bohu" Comparing this with Isaiah 45:18 we read that God "Created it [the earth] not in vain [Tohu] he created it to be inhabited." There certainly appears to be a gap between Genesis 1:1 and 1:2 in which there were beings created by God that inhabited the earth.

## Who lived in the Gap?

As we study Chapter 1, we will see that man was created on the sixth day of creation week. However, when man is created there was one who was there as the man slayer to slay him. In John 8:44 our Lord tells the leaders of Israel that the devil was "…a murderer from the beginning." The word translated "murderer" in John 8:44 is actually the word for "manslayer." In Ezekiel 28:15 we read that Satan was "…Perfect in [his] ways from the day [he] was created." So then he was not a murderer from his beginning. He was however a murderer (a man slayer) from man's beginning. Man was created on the sixth day of creation week. As we study each day of that week we will see that there is no reference for the creation of the angelic world during that week. Therefore we assume that the creation of Lucifer as the anointed Cherub has to have been "in the beginning" and his subsequent fall to become Satan had to have occurred in that gap between Genesis 1:1 and 1:2. We understand too that the events in Isaiah 14:12 – 14 and Ezekiel 28:11-19 (which describes his activities during that time period) have taken place in that gap.

Let's look at the Isaiah 14 and Ezekiel 28 passages to see if we can picture what life on earth was like in that gap. As we study these passages, we do so realizing that there was no human life there. However, we will see that what transpired there would eventually impact man and would also have a great impact in God's eternal purpose for man. Though man was not there, one was there who recorded the events for man's knowledge. These passages are recorded in our Bibles so that we can understand the nature of the spiritual conflict that goes on in the world today and what was the origin of the adversarial relationship that exists today between the purposes of God for man and the efforts by the adversary to prevent God and man from having true fellowship.

### Ezekiel 28 - We meet the Anointed Cherub

[11] Moreover the word of the LORD came unto me, saying, [12] Son of man, take up a lamentation upon the king of Tyrus, and say unto him, Thus saith the Lord GOD; Thou sealest up the sum, full of wisdom, and perfect in beauty. [13] Thou hast been in Eden the garden of God; every precious stone *was* thy covering, the sardius, topaz, and the diamond, the beryl, the onyx, and the jasper, the sapphire, the emerald, and the carbuncle, and gold: the workmanship of thy tabrets and of thy pipes was prepared in thee in the day that thou wast created. [14] Thou *art* the anointed cherub that covereth; and I have set thee *so*: thou wast upon the holy mountain of God; thou hast walked up and down in the midst of the stones of fire. [15] Thou *wast* perfect in thy ways from the day that thou wast created, till iniquity was found in thee. [16] By the multitude of thy merchandise they have filled the midst of thee with violence, and thou hast sinned: therefore I will cast thee as profane out of the mountain of God: and I will destroy thee, O covering cherub, from the midst of the stones of fire. [17] Thine heart was lifted up because of thy beauty, thou hast corrupted thy wisdom by reason of thy brightness: I will cast thee to the ground, I will lay thee before kings, that they may behold thee. [18] Thou hast defiled thy sanctuaries by the multitude of thine iniquities, by the iniquity of thy traffick; therefore will I bring forth a fire from the midst of thee, it shall devour thee, and I will bring thee to ashes upon the earth in the sight of all them that behold thee. [19] All they that know thee among the people shall be astonished at thee: thou shalt be a terror, and never *shalt* thou *be* any more.

**(Ezekiel 28:11-19)**

In Ezekiel 28: 1 -10 which precedes our study passage, we read God's words to one who is called the prince of Tyrus. This person is a man (Ezek. 28: 2 and 9) who lifts himself up as though he were a god. These ten verses are written to a mortal man – the prince of Tyrus. Some Bible students view this as a cryptic message that is prophetic of the coming antichrist who will be a man who claims to be God.  Ezekiel 28: 11 - 19 (our study passage shown above) then switches the focus from the man who is the prince of Tyrus to the spirit being who is called the king of Tyrus. This one is the real power behind the human prince of Tyrus who was lifted up with pride so as to regard himself as God. Let's note what is said of this creature in this passage:

- In verse 12 we read "Thou sealest up the sum…" – He was the top of the hierarchy of God's creation.
- "…full of wisdom…" -- He was a highly intelligent creature.
- "…and perfect in beauty…" -- He was attractive in every way.

- In verse 13 we read "Thou hast been in Eden the garden of God…" -- He was in the geographic area in the Middle East that is called Eden. This does not necessarily refer only to his presence in Eden when Adam and Eve were there. This is probably a reference to his locale when he was first created.

- "…every precious stone *was* thy covering, the sardius, topaz, and the diamond, the beryl, the onyx, and the jasper, the sapphire, the emerald, and the carbuncle, and gold…" He was decked in precious jewels and with gold.

- "…The workmanship of thy tabrets and of thy pipes was prepared in thee in the day that thou wast created." He had musical instruments built into his very being when he was created. This indicates that his role in creation was to be connected with music.

- In verse 14 we read "Thou *art* the anointed cherub that covereth; and I have set thee *so*: …" Cherubim are always seen in connection with the throne of God or the presence of God when they appear in scripture. When Ezekiel saw the throne of God, there were four cherubim seen under and around the throne (Ezekiel 1:4-28; 10:1-22; 11:22); This creature in Ezekiel 28:14 would have been the fifth cherub – the one that would be over the throne as some type of a covering.

- "…Thou wast upon the holy mountain of God;…" This could have been Mount Zion on earth or it could also have been the Mount of the Congregation that Isaiah 14:13 speaks of. It is also possible that this could be referring to the meeting place of the angels to which Job 1:6 and 2:1 refer.

- "Thou hast walked up and down in the midst of the stones of fire." These stones of fire appear to be a reference to the stars. There is some kind of a connection between stars and angels as we see in Revelation 1:20; 2:1; 2:4; 3:1; 8:12; and 9:1. In Isaiah 14:14, Satan said he would ascend above the heights of the clouds. Clouds are also associated in scripture with a great company of angels. It is apparent from this that he desired to rule over the angelic hosts of heaven.

- "Thou *wast* perfect in thy ways from the day that thou wast created, till iniquity was found in thee" (verse 15). This first of all speaks of his creation. He is a created creature. It also tells us that he was not a fallen creature for some time after his creation. There was a point in his career when he fell into sin and iniquity was found in him. This would have happened between his creation in Genesis 1:1 when in the beginning God created the heaven and the earth and the time when the earth became "…without form and void" in Genesis 1:2. Creation week that leads to the creation of man then starts in Genesis 1:3 where we read "And God said…". When man is created in Genesis 1:26 on the sixth day of that week, Satan is there as a man slayer to slay man.

- "By the multitude of thy merchandise they have filled the midst of thee with violence, and thou hast sinned." Here in verse 16 we see that he is marketing a plan that is called "merchandise" that involves violence. We will see his plan in Isaiah 14. In verse 18 we see him trafficking his merchandise.

- "Thine heart was lifted up because of thy beauty…" Here in verse 17 we see the reason for his fall. He was lifted up with pride because of his beauty.

- "…Therefore I will cast thee as profane out of the mountain of God: and I will destroy thee, O covering cherub, from the midst of the stones of fire." This will finally happen in Revelation 12: 7-9 where we will see him being cast out of heaven never to be there again. The Lord sees this event as He looked into the future in Luke 10:18 when he says "…I beheld Satan as lightning fall from heaven." Remember – the Lord declares the end from the beginning. (Isaiah 6:10)

- "Thou hast defiled thy sanctuaries by the multitude of thine iniquities…" (Verse 16). His sanctuary was his place of residence where he was to live. We will see in Isaiah 14 that his residence was to be the earth. What we see in Genesis 1:2 is his place of residence in a defiled state. What God is doing in the creation week of Genesis 1:3 to 1:31 is refurbishing of the earth to make it a suitable habitation for this new creature – man. It was defiled because of a divine judgment on it due to the rebellion in the angelic world.

## Isaiah 14 - Satan's Five Step Plan to take over God's Kingdom

¹² How art thou fallen from heaven, O Lucifer, son of the morning! *how* art thou cut down to the ground, which didst weaken the nations! ¹³ For thou hast said in thine heart, I will ascend into heaven, I will exalt my throne above the stars of God: I will sit also upon the mount of the congregation, in the sides of the north: ¹⁴ I will ascend above the heights of the clouds; I will be like the most High.

(Isaiah 14:12-14)

In Isaiah 14 we see a five point plan by which Satan intended to displace God as the most high that possesses heaven and earth. Let's list the steps in sequence:

1. Step one: "I will ascend into heaven." We can understand from this that he was not supposed to be in heaven. For him the first step in rebellion was that he would ascend into heaven. Remember that in the beginning God created the heaven (singular) and the earth. The heaven would be the stellar universe. The earth would be the earth that we live on today. We understand also that since Lucifer was the anointed cherub that covereth the throne, that God's throne would have been on the earth at that time. There are two species of angelic creatures that are associated with God's throne -- or at least they are around the throne when they are encountered in scripture. They are the cherubim (Isaiah 37:16; Ezekiel 9:3; 10:1-20; Hebrews 9:5) and the seraphim (Isaiah 6:2-6). The throne of God is today in the third heaven (2Corinthians 12:2). A fact to be noted here is that after Genesis 1:1 there is more than one heaven. Apparently God moved His throne off of the earth when sin and rebellion entered into His creation with the defection of Lucifer to become Satan. The first heaven would be the atmosphere of the earth where the birds fly (we will learn more about this in Jeremiah 4:25 later). The second heaven would be where the stars of heaven are (the stellar universe). The third heaven is then actually above the water that was separated from off of the earth on the second day of creation week to create the open firmament of heaven. The deep is still there but it is split so that part of it is on the earth and the other part is the water separating the second heaven (the stellar universe) from the third heaven where the throne of God is today. Job 38:30 says of it that the surface of the deep is frozen. This would be what John saw in Revelation 4:6 as the floor of the third heaven which he described as a sea of glass like unto crystal. God created a third heaven as a temporary residence when He removed His throne from the earth when sin entered through the fall of Satan and the rebellion in the angelic world. Ultimately, God's residence will be the New Jerusalem of which Revelation 21:3 speaks. See the book "*A Study in the Book of the Revelation*" by the same author.

2. Step 2 "I will exalt my throne above the stars of God." Satan has a throne and he intended to exalt it above the stars of God. We understand that these stars of God are references to angels as there is some Bible connection of angels with stars. We see this connection in Revelation 12:4 where the tail of the great red dragon (who we understand to be Satan) drew a third of the stars of heaven and cast them to the earth. We take that to mean that a third of the angels had followed Satan in rebellion.

3. Step 3 "I will sit also upon the mount of the congregation, in the sides of the north." The Mount of the Congregation is apparently the meeting place in the universe where angels meet to carry on the affairs of heaven much as how men meet on earth to do that in the world of men. Satan's plan here is to rule over the angelic world. He is today the "prince [arche in the Greek] of the power [government] of the air [the unseen realm of heaven]" (Ephesians 2: 2) and also "the god of this

world" (2Corinthians 4:4). Both realms (the heaven and the earth) are created by Christ and for Christ (Colossians 1:16) but both are today still in the hands of the usurper today.

4. Step 4 "I will ascend above the heights of the clouds." Here again the clouds are a great company of angels (Matthew 24:38; 26:64; Revelation 1:6). Satan planned to rule over the world of angels. Remember man was not yet created at that time.

5. Step 5 "I will be like the most High." This term is found in Genesis 14:18 and 19 where we see Melchizedek who was a priest of the most high God say to Abram "Blessed be Abram of the most high God, possessor of heaven and earth." Satan intended to take from his creator the title "the Most High God" and in doing so to be "…the possessor of heaven and earth."

All of this happened between Genesis 1:1 (the original creation of heaven and earth when the angelic world was created) and Genesis 1:2 when we see the earth under a divine judgment because of the sin and rebellion that occurred in the angelic world that existed on earth at that time. We can get a picture of the judgment that came on the earth to leave it in the state in which we find it in Genesis 1:2 by a study of the passage in Jeremiah 4:23-26. The Jeremiah passage seems to reflect back on Genesis 1:2 in that it describes the state of the earth as it was in at the start of the six days of what we call creation week.

What we find in Jeremiah Chapter 4 is the Lord lamenting the judgment that He will have to bring upon Judah and Jerusalem because of the sin and rebellion of its inhabitants. It is the Babylonian captivity of Judah that is in view in Jeremiah that will result in the land being empty. However, the Lord is reflecting back on an earlier time when the whole earth was emptied of its inhabitants as a result of sin and rebellion.

## Jeremiah 4 - How the Earth became Without Form and Void

> 23 I beheld the earth, and, lo, *it was* without form, and void; and the heavens, and they *had* no light. 24 I beheld the mountains, and, lo, they trembled, and all the hills moved lightly. 25 I beheld, and, lo, *there was* no man, and all the birds of the heavens were fled. 26 I beheld, and, lo, the fruitful place *was* a wilderness, and all the cities thereof were broken down at the presence of the LORD, *and* by his fierce anger.
>
> (Jeremiah 4:23-26)

Let's study this passage as a reflection back on the desolation of the earth at the close of that gap between the original creation of the heaven and the earth and the re-creation (the remaking and repurposing) of the earth to make it a suitable habitation for man.

- "I beheld the earth, and, lo, *it was* without form and void; and the heavens, and they *had* no light." The stars were there but the sun and moon were not. This is exactly how the Spirit of God found the earth when He started to move upon the face of the waters in Genesis 1:2.
- "I beheld the mountains, and, lo, they trembled, and all the hills moved lightly." This did not happen to the land of Judah during the Babylonian invasion and the subsequent captivity of the nation of Israel. This in Genesis 1:2 and Jeremiah 4:23 was a severe judgment that totally shook the earth and left it under water before Genesis 1:2.
- "I beheld, and, lo, *there was* no man…" This too is the state of affairs in Genesis 1:2. Man would not be created yet for another six days. It was not however, the state of affairs for the land of Palestine during the Babylonian captivity. The poorest of the people were still left in the land yet during the captivity.
- "And all the birds of the heavens were fled…:" This would not have happened in the Babylonian captivity. Note that it refers to the birds of the heavens (plural). There are more birds of the heavens than the winged creatures that we know as birds today. There were the cherubim and seraphim in the earth if the throne of God was there. In Revelation 18:2 we see the site of Babylon after its

ultimate destruction "…become the habitation of devils, and the hold of every foul spirit, and a cage of every unclean and hateful bird." This would not describe birds that we know today. These would be those of the seraphim and cherubim who defected with Lucifer's rebellion.

People who oppose the concept of a gap existing between Genesis 1:1 and 1:2 do so by citing Exodus 20:11. Let's consider this passage:

> "For *in* six days the LORD made heaven and earth, the sea, and all that in them *is*, and rested the seventh day: wherefore the LORD blessed the Sabbath day, and hallowed it."

To understand this verse in light of the gap, we must recognize that there is a difference between what is "created" and what is "made." God created things by bringing something into existence that did not exist before God's act of creating it. Things that are made however were made by God from pre-existing materials – materials that were created in Genesis 1:1. We will study the difference between these two acts of God later.

Other people raise a criticism of the gap concept by suggesting that it is used to accommodate dinosaurs. That is not what we are proposing here in our study of Genesis. There were two notable dinosaurs that are mentioned in the Book of Job that were created on the sixth day with Adam. Behemoth is mentioned in Job 40:15 and Leviathan in Job 41:1 and other Old Testament passages.

## Can We Date Creation?

Appendix 1 of this Study in Genesis presents a Framework for Dating Creation Based on information in the Bible. However, it must be noted that this framework is limited to the dating of what we are calling Creation Week – the six day period of time covered in Genesis 1 verses 3 through 31. There is sufficient information that the date for the creation of Adam and Eve can be calculated fairly accurately to within about 25 years of the exact date. Illustration 1 in Appendix 1 is a time line illustration that presents estimates of these dates. The dates in this illustration are based on dating contained in the book *The Chronology of the Bible* by Frank R. Klassen (copyright 1975) Published by Regal Publishers, Inc. I use his chronology in Illustration 1 because it fits in closely with the majority of others (see the table of other dates in Appendix 1) who have made exhaustive effort to calculate that date from the Bible and it fits closely to my personal study. The dates in Illustration 1 are not intended to be the final word on the exact date of creation week but can be used to show the relative dates of biblical and historical events as to how they relate to each other in time.

It must be remembered in our quest to estimate the age of the earth that we do not have information in the pages of scripture that will enable us to date the original creation of the heaven and the earth. What we are to understand is simply that at some point in the distant past, God created the heaven and the earth as stated in Genesis 1:1. At that original creation, the entire space-matter-time continuum was brought into existence instantly all at the same time. Time and Matter could not have been created without Space or there would be no space in which it could happen. Space and Time could not have been created without Matter or there would have been nothing that was created. Matter and Space could not have been created without time or there would have been no time in which it could have happened. When the space-matter-time continuum was created, it had to have been created by an all powerful God who existed outside of the space-matter-time universe. The universe could not have created itself just as surely as a highly complex computer could not have created itself. There had to have been an omnipotent, omniscient, omnipresent, eternal God who existed outside of and independent of creation to do the work of creation.

Understanding the laws of science that govern the universe (notably the Second Law of Thermodynamics), we understand that the space-matter-time continuum could not have always existed. The second law of thermodynamics basically says that in any energy transformation process, a certain amount of energy goes into its waste form – that being heat. If the universe had been here for an infinite period of time all of the energy in the universe would have gone into its waste form of heat. At that point, the universe would have died a heat death and the entire universe would be at a uniform and constant temperature. Also at that point, no further motion could be possible because energy has to be at two different levels (two different temperatures) for motion to occur.

The Second Law of Thermodynamics (what we call the decay principle) possibly could have then been put in place as part of the curse that God put on His creation as a result of Adam's sin in Genesis Chapter 3. We will study more on this later. Further, we understand that the space-matter-time continuum could not have come into existence spontaneously of itself as that would violate the first law of science (the first Law of Thermodynamics) which says that matter and energy can neither be created nor destroyed. From this we learn and understand that during the process of creation the laws that govern the present universe were not in effect at the time of creation. They were laws that God put in place later to govern His creation after he created it.

## How Old is the Universe?

Perhaps a better question is "Do we need to know how old the universe is?" We do not know nor can we know the age of the universe. We also do not know nor do we have information on the chronology of exactly what transpired between Genesis 1:1 and 1:3. That information is hid from us. God knows it but He has chosen to not reveal that to us mortal men. We do know from scripture of certain events that took place during that time period but we do not have definitive information of the dates or sequence of events. We can speculate on it but that is just that – speculation.

We know from Isaiah 14 that Lucifer fell from his position as the anointed Cherub during that time period. On the one hand we can speculate that it did not take him long to be lifted up with pride and decide that he should be the one in charge and took action to organize an insurrection to take over. On the other hand, as we study the universe we see that it looks old and in fact the Bible gives indications that it is old as we read the many verses such as Second Peter 3:5 and 6 "For this they willingly are ignorant of, that by the word of God the heavens were of old, and the earth standing out of the water and in the water: Whereby the world that then was, being overflowed with water, perished:…" Also there are other verses that speak of the universe being old: Psalm 68:33 "To him that rideth upon the heavens of heavens, *which were* of old; lo, he doth send out his voice, *and that* a mighty voice." We see that term "Of old…" repeated over and over in the pages of scripture with reference to God's work in the universe.

We can put a date on Genesis 1:5 but we can't on Genesis 1:1 and it is certainly not necessary that we do. Those who are trying desperately to do so are basing their premise on the supposition that the stellar universe had to have been created in day four of creation week. The whole starlight and time question comes down to this: "If the universe is about 6,000 years old, how is it that we see light from stars that are several billion light years away?" The premise of the universe being only 6,000 years old is a real hurdle in light of what we actually know about the universe. There are two Bible verses that we want to look at in some detail and then we will look at other Bible verses that can answer the perplexing question regarding the age of the universe itself and yet taking a stand for the literal six day creation week of Genesis Chapter 1. It is this author's opinion that the universe is probably less than 10,000 years old in earth years. It is also this author's confidence that creation week as we find it in Genesis Chapter 1 is a literal week of seven days of 24 hours each.

Let's look first at Genesis 1:6-8 regarding the "making" of the open firmament of heaven in which He will put the sun and the moon.

> Genesis 1:6-8 "⁶ And God said, Let there be a firmament in the midst of the waters, and let it divide the waters from the waters. ⁷ And God made the firmament, and divided the waters

which *were* under the firmament from the waters which *were* above the firmament: and it was so. ⁸ And God called the firmament Heaven. And the evening and the morning were the second day."

We will look at the passage in Genesis 1:14 – 19 in which we see the "making" of the sun and the moon. Remember that to "make" something is to fashion it (to produce it) using previously existing material.

Genesis 1:14-19 "¹⁴ And God said, Let there be lights in the firmament of the heaven to divide the day from the night; and let them be for signs, and for seasons, and for days, and years: ¹⁵ And let them be for lights in the firmament of the heaven to give light upon the earth: and it was so. ¹⁶ And God made two great lights; the greater light to rule the day, and the lesser light to rule the night: *he made* the stars also. ¹⁷ And God set them in the firmament of the heaven to give light upon the earth, ¹⁸ And to rule over the day and over the night, and to divide the light from the darkness: and God saw that *it was* good. ¹⁹ And the evening and the morning were the fourth day. "

In Genesis 1:16 we see that God made two great lights; the greater light to rule the day [that would be the sun] and the lesser light to rule the night [that being the moon]. Then the passage says "...*he made* the stars also..." We note that the words "*he made*..." are in italics in the KJV text. That means that they were added by the translators. Note also that the term for God's action is that He "made"_the two great lights - the sun and the moon. While the action of creating something is to bring something into existence that was not there before, the action defined by the term "made" has to do with making something out of previously existing material. The sun and the moon were both "made" on the fourth day of creation week out of previously existing material. The term "*he made* the stars also..." is a parenthical expression which carries the thought that the stars rule the night together with the moon. They (the stars) were created in the beginning as stated in Genesis 1:1 to comprise the heaven. They were now, with the revamping of the earth in preparation for man's occupancy of it, to rule the night along with the moon. The sun, the moon, and the stars would "...be for signs, and for seasons, and for days, and years:..." and "to give light upon the earth." (Genesis 1:16 – 17)

Another Bible passage that we need to study in light of creation week and Genesis 1:1 is Exodus 20:11. That verse says: "For *in* six days the LORD made heaven and earth, the sea, and all that in them *is*, and rested the seventh day: wherefore the LORD blessed the sabbath day, and hallowed it." Here again the word "made" rather than create is used since it is concerned with the making of the heaven and the earth as we see it today. Also, the word "heaven" is not always a reference to the stellar universe. The word heaven is used 88 times in the first five books of the Bible but about half of those references have to do with the atmosphere and the air. In this passage, "heaven" has reference to anything above the earth's surface. Genesis 1:8 defines the word "Firmament" as the open space where the sun and moon would be.

## The Question of Starlight and Time

Another question that perplexes Bible believers but should not and need not trouble them is the fact that astronomers basically hold that the universe must be at least 13.8 billion years old. Astronomers tell us that the most distant galaxies viewed by the Hubble telescope are 46 billion light years away. This is the case when viewing in all directions. A Light Year is a measure of distance – that being the distance that light travels in one year. That would make the visible universe to be 92 billion light years in diameter. That is huge and that is just the visible universe. There is more of the universe beyond that which is not visible. It is not visible but they know it is there because they pick up infrared radiation, microwaves and radio waves from it using specialized equipment and antennas. The oldest thing in the universe that we can see according to astronomers is the Cosmic Microwave Background (CMB). As objects move away from the earth -- or more properly (as we shall see later) have moved away from the earth in the past, they get too far away for their light to be seen from earth. However, astronomers can pick up the residual radiation from them. At that point, they can only be seen with microwave antennas which pick up their Cosmic Microwave Radiation Background (CMRB) which is left of the stars that disappear from view.

All of this information is based on assumptions that are taken for granted but are not proven. Some of those assumptions are:

1. The speed of light is constant and has always been the same. Actually the speed of light measurements taken over the past 70 years show that the speed of light values could fit on a typical decay curve. However there are questions on the accuracy of the earlier measurements. For the past 40 years the speed of light has been consistently measured as the commonly accepted speed of about 186,282 miles per second. However, these measurements have all been taken using an atomic clock that was invented in 1956. The atomic clock is based on the wavelength of the cesium 133 atom which travels at the speed of light. Therefore, if the speed of light is slowing down, so is the measurement of it. It should be remembered that just as the First Law of Thermodynamics was not in play during the act of creation, other factors that define the present earth such as the speed of light might have been different at that time as well. We will look at this in more detail later.

2. We do not know for certain exactly what light really is. Is it a wave, a photon, or is it a particle? Astronomers assume that it is a wave and use that as a means of analysis by the Doppler Effect on the color spectrum to determine the blue shift or red shift for the determination of the direction of the movement of stars and galaxies and the speed at which they move.

3. The method of measuring distance to stars is very limited as to distance of measurements and accuracy. The basic method of measurement in the past was with what is called Parallax Trigonometric Measurement. This is using basic trigonometry in which if you know two angles and one side of a triangle you can figure out all of the other measurements. The diameter of the earth's orbit is one measure and the deflection angle for the same point on the earth taken to the same point on an object in space at six month interval gives the other two measurements. The more recent procedures utilize the Space Interferometer Mission (SIM) methods. This is said to be able to measure objects that are out 82,000 light years in distance.

4. It is standard procedure in using Global Positioning Satellites (GPS) that the time at the altitude of the satellites is traveling faster than time on earth. Clocks in space actually tick faster than clocks on earth. This phenomenon is demonstrated by use of two atomic clocks on earth with one at sea level at Greenwich England and the other at an elevation of about 5,000 feet above sea level at Bolder Colorado. The atomic clock at 5,000 feet will actually show 5 nanoseconds per year gain over the one at sea level. With that fact, we do not know how fast a clock at a distance of a million or a billion light years or more in space would tick compared to one on earth.

5. Astronomers assume that the universe is currently still expanding. They base this on the fact of the red shift in the color spectrum from stars and galaxies. However, this red shift can also indicate that the universe had expanded in the past but it does not necessarily indicate that it is currently expanding. We will see more on this in the discussion below.

**The Hubble Telescope**

What astronomers are finding with the Hubble telescope is interesting. They found:

- The light from most of the objects in the universe (notably the stars and the galaxies) is shifted to the red end of the color spectrum. The wave length of light from objects that are moving away from earth (or had moved away in the past) is shifted to the red end of the spectrum while light from objects that are moving towards us (or had moved towards us) would be shifted to the blue end of the spectrum. What they see and observe though, is that the light from essentially all of objects in the universe is shifted to the red end and thus we understand that they are moving away from us or had moved away in the past. This also tells us that the earth is at or very near the center of the universe.

- This phenomenon might suggest to us that the universe is expanding or more likely has expanded in the past but that expansion has been stopped. Astronomers also tell us based on the red shift phenomenon that objects that are farther out are apparently moving away from us at an accelerating rate. That is, they are (or appear to be) moving away from earth at a faster rate than closer objects.

- This, according to astronomers, also suggests that the universe is expanding. We will therefore (if this is true) be constantly losing galaxies from the visible universe as they pass out of the visible sphere of the universe. Though astronomers hold that this phenomenon indicates that the universe is currently expanding, we will see that this could also indicate that the universe has expanded in the past but has since stopped expanding (or has been stopped from expanding).

## There is a Biblical answer to the Starlight and Time issue

The fact that we can see stars (and galaxies) that are millions of light years away can be seen and understood by verses such as Psalm 104:2. I quote it here with its Bible context:

> Psalm 104:1-6 (KJV)
> [1] Bless the LORD, O my soul. O LORD my God, thou art very great; thou art clothed with honour and majesty. [2] Who coverest *thyself* with light as *with* a garment: who stretchest out the heavens like a curtain: [3] Who layeth the beams of his chambers in the waters: who maketh the clouds his chariot: who walketh upon the wings of the wind: [4] Who maketh his angels spirits; his ministers a flaming fire: [5] *Who* laid the foundations of the earth, *that* it should not be removed for ever. [6] Thou coveredst it with the deep as *with* a garment: the waters stood above the mountains.

Other verses that speak of God stretching out the heavens as a curtain include Isaiah 40:22; 42:5; 45: 12; Jeremiah 10:12 and 51:15. This gives us insight into how God created the heaven and the earth in Genesis 1:1 so that the light was already visible on earth immediately after creation in spite of the fact that they might be millions of light years away. The stars and galaxies would have been at or near the earth at the center of the universe when God created them. God then moved them out to the position that they currently hold in space. The question then is not how did the light from those stars reach us so fast when they are so far away but rather how did God move those stars so far out into space so fast. These heavenly objects would have had to be moving at a speed greater than the speed of light as it is today and yet leave a trail of light as they are moved away from the earth. Remember, the First and the Second Laws of Thermodynamics would not have been in play at the time of the original creation, it is likely that some of the physical constants such as the current speed of light was not what it is today either.

In an expanding universe all of the heavenly bodies would have moved away from each other as God stretched out the heavens much like raisins move away from each other in bread dough as the bread dough rises. As the stars were moved out, the trail of light emitted from them was left to provide light upon the earth when God moved them into their place in the heaven. Remember the purpose for the sun, moon and the stars was "to give light upon the earth" (Genesis 1:15, 17).

Let's ask at this point: "Is the universe still expanding?" Based on what we see in Psalm 104:1-5, it is clear that the universe had expanded by a great amount in the past. The red shift from the light spectrum from the stars and galaxies also verifies that the universe had expanded. We also understand from Isaiah 14 and Ezekiel 28 that God's throne was once on the earth. That would have been between Genesis 1:1 when He created the heaven and the earth and Genesis 1:3 when He started the six days of creation week. However, from 2Corinthians 12:1 and 2 we know that the throne of God is today in the third heaven.. So then God had moved His throne from off of the earth to the third heaven at some point in the past. We will look at these verses later in our study, but let's consider Job 37:18 "Hast thou with him spread out the sky, *which is* strong, *and* as a molten looking glass?" Let's consider also Revelation 4:6 and 15:2 which tell us what John saw when he saw the throne of God in the third heaven.

## Revelation 4:6 (KJV)

[6] And before the throne *there was* a sea of glass like unto crystal: and in the midst of the throne, and round about the throne, *were* four beasts full of eyes before and behind.

## Revelation 15:2 (KJV)

[2] And I saw as it were a sea of glass mingled with fire: and them that had gotten the victory over the beast, and over his image, and over his mark, *and* over the number of his name, stand on the sea of glass, having the harps of God.

This sea of glass would be the frozen surface of the deep that Job 38:30 speaks about. It is also what Job 37:18 talks about when it refers to the sky being "...strong as a molten looking glass." All of this would tend to indicate that God stopped the expansion of the universe at some point in the past and moved His throne from the earth to the third heaven. When He did that, He put a boundary on the universe and separated the second heaven from the third heaven by the frozen face of the deep. This would have happened before He inundated the earth with the deep that Genesis 1:2 talks about.

## A Stable Universe

Until recently, cosmologists regarded the universe as being stable – it is neither expanding nor contracting. As noted above, the universal shift of the color of star light to the red end of the spectrum tells us two things regarding the created universe: 1) the universe had expanded in the past, and 2) we on earth are at or near the center of the universe. The information regarding the sky being "...strong as a molten looking glass" in Job 37:18 and the floor of the throne room of God in the third heaven as we saw in Revelation 4:6 and 15:2 suggests that the universe is still stable and is not expanding. The red shift of starlight would then be the residual effect of the stretching out of the heavens that the passage in Psalm 104:1-6 speaks of.

For the universe to be stable there would have to be two forces in play. One force would be the gravitational forces that would tend to make the universe collapse on itself to its center of mass – that being the earth or our solar system or perhaps the milky weigh galaxy. The other force would then be the counterforce which would prevent that from happening. Astronomers tell us that there is a force in the universe that they estimate accounts for up to 85% of its mass. They can not detect it or measure it but they feel confident that it is there. For lack of better terms, they call it dark matter or dark energy. This dark energy would then be the force that counters the forces that would collapse the universe to its center of mass.

This balancing of forces is a part of what the apostle Paul refers to in Colossians 1: 16-17. "[16] For by him were all things created, that are in heaven, and that are in earth, visible and invisible, whether *they be* thrones, or dominions, or principalities, or powers: all things were created by him, and for him: [17] And he is before all things, and by him all things consist." The word translated "consists" is from a Greek word meaning "to cause to stand together." This is talking about our Lord Jesus Christ as the creator of all things and carries the thought that not only is He the creator of all things, but He is the one who holds everything together – including our physical universe and everything therein. Even the molecular structure of the very cross that He died on in order to accomplish redemption for us was held together by the power of His will. What a loving and wonderful Savior that we have!

## What was here before Genesis 1:1?

One question that comes to mind as we study God's act of creating everything that has been created, is "What was here before Genesis 1:1" when "In the beginning God created the heaven and the earth?" The answer to that question is that the eternal, self existing, omnipresent, omnipotent, and omniscient and purposeful God existed in the three separate and distinct persons that we now know Him from scripture as the Father, the Word and the Holy Spirit. They had and continue to have their existence outside of the space-matter-time continuum that we know as the created universe of which we are a part. We in our finite minds that are locked into space, matter, and time have great difficulty relating to that eternal existence of God. Psalm 90 verse 2 states the matter well saying: "Before the mountains were brought forth, or ever thou hadst formed the earth and the world, even from everlasting to everlasting, thou art God." But Jesus, the eternal Word, entered into the space-matter-time continuum that He created to redeem man.

There are passages that take us back there to give us a view of what life in and among the trinity of God (what the KJV Bible calls the "Godhead") was like in passages as John 17:5-8 where we find the Son speaking to the Father about this eternal existence:

> "[5] And now, O Father, glorify thou me with thine own self with the glory which I had with thee before the world was. [6] I have manifested thy name unto the men which thou gavest me out of the world: thine they were, and thou gavest them me; and they have kept thy word. [7] Now they have known that all things whatsoever thou hast given me are of thee. [8] For I have given unto them the words which thou gavest me; and they have received them, and have known surely that I came out from thee, and they have believed that thou didst send me. (John 17:5-8)

We see into the past from John 1:1-4 that before time even began there was God. But we see from this verse that there are two persons both of whom are called God in this amazing passage. There is the one who is called "...the Word" who was there with God and "...the Word was God." There are two persons in this passage both called God. There is God (who we would know as the Father) called God but there is one who is called the Word who is also called God. In verse 14 of John Chapter one, we see that the Word was made flesh . We understand from this passage that the eternal Word is Jesus who became a man as the only begotten Son of God.

We see this again in the dialogue that Jesus had with the scribes and Pharisees in John Chapter 8 with the words "...before Abraham was, I am." This is a claim to be the self existing God. So we see from this that Jesus is God but we find also in scripture that there is one called the "eternal Spirit" (Isaiah 51:15 and Hebrews 9:14) who is also called God (Acts 5:4).

Before time began there was God and He, in the perfect fellowship of the trinity of the Godhead had no need for anyone else to have perfect harmony of purpose and complete fellowship. They had no need of anyone else for personal fulfillment. Yet God decided to create a universe in which He would place free, moral agents with whom He can have fellowship. Those free moral agents included us members of this human race along with angels, saraphim, and cherubim.

## Man created in the Image and Likeness of God

What we find in the Book of Genesis is the work of God creating man in His own image. Consider the words of God regarding the creation of man in Genesis 1:26-27 .

> "And God said, Let us make man in our image, after our likeness: and let them have dominion over the fish of the sea, and over the fowl of the air, and over the cattle, and over all the earth, and over every creeping thing that creepeth upon the earth. So God created man in his own image, in the image of God created he him; male and female created he them." (Genesis 1:26-27)

If we are to understand God, we need to look at our own makeup as men because we are created in God's image. I present below an excerpt from the book *"You and your Creator"* (Tiry, 2021) to make this parallel in order to illustrate who God is and how we are to relate to Him.

## From the book *"You and Your Creator"*

The Bible says that there is one God and yet in the Bible there are three persons in the Bible called God. Each has their own personality. Each operate independently of each other but when they act they do so in perfect harmony of purpose with each other. God is called "the Father" in the Bible. (Romans 1:7) He is called the Father because He is the Father of our Lord Jesus Christ (Romans 15:6). He has a Son (Proverbs 30:4) who He (the Father) also calls God: "But unto the Son *he saith*, Thy throne, O God, *is* for ever and ever: a sceptre of righteousness *is* the sceptre of thy kingdom. Thou hast loved righteousness, and hated iniquity; therefore God, *even* thy God, hath anointed thee with the oil of gladness above thy fellows. And, Thou, Lord, in the beginning hast laid the foundation of the earth; and the heavens are the works of thine hands:" (Hebrews 1:8 - 10).

The one who the Father calls the Son is the eternal Word from eternity past. We find the reference to the eternal Word in John 1:1-4 where we read "In the beginning was the Word, and the Word was with God, and the Word was God. The same was in the beginning with God. All things were made by him; and without him was not any thing made that was made. In him was life; and the life was the light of men."

This in John 1:1-4 is an amazing passage. In the beginning there was a person called the Word and that person was with God and further, the person who is called the Word was God. There are two persons in this one verse both of whom are called God. It is particularly interesting to note from this verse that everything that has been made was made by Him (the Word). This excludes Himself so that we are to understand that the Word eternally existed with the Father from eternity past.

In Colossians we see another intensely fascinating passage regarding this one who is called the Son. Let's look carefully at this passage because it is packed with rich and wonderful information about who the Son is and about the relationship that we as believers have with the Father through the Son.

> "[12] Giving thanks unto the Father, which hath made us meet to be partakers of the inheritance of the saints in light: [13] Who hath delivered us from the power of darkness, and hath translated *us* into the kingdom of his dear Son: [14] In whom we have redemption through his blood, *even* the forgiveness of sins: [15] Who is the image of the invisible God, the firstborn of every creature: [16] For by him were all things created, that are in heaven, and that are in earth, visible

and invisible, whether *they be* thrones, or dominions, or principalities, or powers: all things were created by him, and for him: [17] And he is before all things, and by him all things consist. [18] And he is the head of the body, the church: who is the beginning, the firstborn from the dead; that in all *things* he might have the preeminence. [19] For it pleased *the Father* that in him should all fulness dwell; [20] And, having made peace through the blood of his cross, by him to reconcile all things unto himself; by him, *I say*, whether *they be* things in earth, or things in heaven." (Colossians 1:12-20)

## The Son is called God

This passage in Colossians Chapter 1 speaks volumes about the nature of the relationship that believers have with the Father through the Son. First of all, the Father made the believer fit (adequate, suitable) to be a partaker of the inheritance of the saints. He did this through the work that the Son did for us. The Father delivered the believer from the power of darkness and put him into the kingdom of His dear Son. We therefore understand that the Son has a kingdom and the believer is placed into it. The believer has redemption through the blood of the Son and therefore has the forgiveness of sins. To have redemption is to be purchased from someone or something (which in this passage is the power of darkness). The price of that redemption is the blood of the Son. This is the blood that Jesus Christ, the Son shed for man on Calvary. He surrendered His life (the only perfect human life ever lived) as full payment for man's sins.

In verse 15 of Colossians 1 we see that He (the Son) is the image of the invisible God, the firstborn of every creature. He is the visible manifestation of the God that otherwise could not be seen. He is the first born of every creature because every creature was created by Him and for Him (see also Hebrews 2:10 and Revelation 4:11 on this). The reference to "first born" is a reference to the fact that He (the Son) inherits everything. He inherits everything because He (the Son) created everything. John 1:10-14 states it saying ""[10] He was in the world, and the world was made by him, and the world knew him not. [11] He came unto his own, and his own received him not. [12] But as many as received him, to them gave he power to become the sons of God, *even to* them that believe on his name: [13] Which were born, not of blood, nor of the will of the flesh, nor of the will of man, but of God. [14] And the Word was made flesh, and dwelt among us, (and we beheld his glory, the glory as of the only begotten of the Father,) full of grace and truth."

## The Holy Ghost is called God

In Acts 5:4 we meet one called the Holy Ghost and find that he too is called God. This brings us to the concept of the Trinity. People who oppose the doctrine of the trinity readily point out that the word "Trinity" is not even in the Bible. That is true but the doctrine is clearly there. The Bible word for the Trinity of God is the word "Godhead" as we see it in Acts 17:29; Romans 1:20; and Colossians 2:9. Doctrinal statements regarding the Godhead basically state that God is one in essence and being but three in person. Seeing then the Trinity of God and understanding that God created man in His image, we expect to see a Trinitarian makeup to man. Here again we go to the Word of God and indeed do find such a concept in 1Thessalonians 5:23 where the apostle prays "And the very God of peace sanctify you wholly; and *I pray God* your whole spirit and soul and body be preserved blameless unto the coming of our Lord Jesus Christ."

Man is created as a three part creature. Unlike the creator, man is not three in person as the creator is. Man individually is singular in personality. There is however only one humanity created by God which encompasses many persons who comprise this human race (and there is only one race -- that being the human race composed of billions of individuals).

## Man Created in the image and likeness of His Creator

It is the soul of man that constitutes the person of the individual man. However, while each soul is a unique person, each of the three parts of man has what we would call a mentality that can each be identified in the Bible. The spirit of man has a mentality that the Bible calls "the mind." It is there in our human spirit where we formulate the paradigms by which we interact with each other and with God. The human spirit is that part of our makeup that is given by God to enable us to relate to God and to each other. There is a correspondence between the human spirit and the Spirit of God. We see this in passages as Romans 8:15 - 17 "For ye have not received the spirit of bondage again to fear; but ye have received the Spirit of adoption, whereby we cry, Abba, Father. The Spirit itself beareth witness with our spirit, that we are the children of God: And if children, then heirs; heirs of God, and joint-heirs with Christ; if so be that we suffer with *him*, that we may be also glorified together."

The soul of man has a mentality that the Bible calls "the heart." It is there in the heart (in the mentality of our soul) that we form our affections. It is there where we make decisions in life. It is the heart that believes as the apostle says in Romans 10:9 and 10 "That if thou shalt confess with thy mouth the Lord Jesus, and shalt believe in thine heart that God hath raised him from the dead, thou shalt be saved. For with the heart man believeth unto righteousness; and with the mouth confession is made unto salvation."

If the spirit of man corresponds to the Holy Spirit of God, then there is a correspondence of the soul of man with the Father. First Corinthians 6:19-20 says "What? know ye not that your body is the temple of the Holy Ghost *which is* in you, which ye have of God, and ye are not your own?For ye are bought with a price: therefore glorify God in your body, and in your spirit, which are God's." Verse 20 is a direct command. We understand from grammar that the subject of a direct command is "you". The "you" here is the "you" of your soul. You are a soul (an individual person) which has a body and which also has a spirit and you (the soul that is you) decides what you will do with each.

The soul is then the decision making part of our makeup. There appears then to be a correspondence between the souls of man in the tri-part makeup of man to the Father in the Godhead. As we study our Lord Jesus Christ in His relationship with the Father we find that He lived His entire life on earth in full reliance and dependence upon the Father. Note the Lord's words in John 5:30 "I can of mine own self do nothing: as I hear, I judge: and my judgment is just; because I seek not mine own will, but the will of the Father which hath sent me." Jesus Christ as the eternal Word is sent into the human race by the Father to be fully man while yet being fully God. He does that to provide a way for man to have eternal life. Note the Lord's word in John 5:36-40 to the religious leaders of Israel "[36] But I have greater witness than *that* of John: for the works which the Father hath given me to finish, the same works that I do, bear witness of me, that the Father hath sent me. [37] And the Father himself, which hath sent me, hath borne witness of me. Ye have neither heard his voice at any time, nor seen his shape. [38] And ye have not his word abiding in you: for whom he hath sent, him ye believe not. [9] Search the scriptures; for in them ye think ye have eternal life: and they are they which testify of me. [40] And ye will not come to me, that ye might have life."

Let's review what we see so far regarding the Lord Jesus Christ with regard to creation. We see that He created everything that has been created (Colossians 1:16). We saw also that everything that has been created was created by Him and for Him. When God interacts with creation, it is the eternal Word who does it. He then is the visible manifestation of the God who is otherwise invisible. We see this in John 1:14 where we read "And the Word was made flesh, and dwelt among us, (and we beheld his glory, the glory as of the only begotten of the Father,) full of grace and truth."

Interestingly no one has seen me and neither has anyone seen you. People see the physical manifestation of you but the real you (the soul that is you) no one has seen. It is through your physical body that there is a physical manifestation of you. Now with regard to the tri-part makeup of man, it is through our physical body that we interact with the physical creation. There is then a correspondence between our physical bodies and the Lord Jesus Christ. While we have physical life, we are able to live in God's physical creation which was created by Jesus Christ, the eternal Word.

The created universe itself is a trinity. It is a space-matter-time continuum. Each of these is Trinitarian in nature. Space has three dimensions of length, width and depth. Space would not exist without all three. Matter has three states of being – solid, liquid, and gas. All three are necessary for the existence of matter. Time has three tenses of past, present and future. If any of these were removed, there would not be time.

At this point I direct the reader's attention to Figure 1. This figure illustrates in graphical form information presented in the narrative presented above. You will notice that the Godhead exists outside of and independent from His creation. You will also notice that creation (the universe) has a Trinitarian makeup in its entirety. Note that man is created in God's universe. Also, notice that the Word is the one who interacts with God's creation. The Word enters His creation as the only begotten Son to accomplish redemption by His work of redemption on Calvary. Notice too that the Holy Ghost then administers the redemption program on behalf of the Godhead. The Holy Ghost works in response to the faith of the believer to accomplish salvation of the sinner's soul. He does that by going into that person's life and bundles up the sin and all uncleanness and imputes them to the account of the Son on Calvary. The redeeming work of Christ then clears the account of the sinner's debt by virtue of His death for our sins (1Corinthians 15:1-4). The Father then receives that person into His family as the sinner trusts that work of redemption (Eph. 1:6).

## God had an Eternal Purpose in the Creation of Man

God had an eternal purpose in mind when He created man in His own image. However, He did not state that purpose up front when He created man because there was a reason for keeping it secret. The apostle Paul makes reference to that eternal purpose in Ephesians 3: 1-13 "[1] For this cause I Paul, the prisoner of Jesus Christ for you Gentiles, [2] If ye have heard of the dispensation of the grace of God which is given me to you-ward: [3] How that by revelation he made known unto me the mystery; (as I wrote afore in few words, [4] Whereby, when ye read, ye may understand my knowledge in the mystery of Christ) [5] Which in other ages was not made known unto the sons of men, as it is now revealed unto his holy apostles and prophets by the Spirit; [6] That the Gentiles should be fellow-heirs, and of the same body, and partakers of his promise in Christ by the gospel: [7] Whereof I was made a minister, according to the gift of the grace of God given unto me by the effectual working of his power. [8] Unto me, who am less than the least of all saints, is this grace given, that I should preach among the Gentiles the unsearchable riches of Christ; [9] And to make all *men* see what *is* the fellowship of the mystery, which from the beginning of the world hath been hid in God, who created all things by Jesus Christ: [10] To the intent that now unto the principalities and powers in heavenly *places* might be known by the church the manifold wisdom of God, [11] According to the eternal purpose which he purposed in Christ Jesus our Lord: [12] In whom we have boldness and access with confidence by the faith of him. [13] Wherefore I desire that ye faint not at my tribulations for you, which is your glory. "

God has an eternal purpose for creating man. Though He created man after the creation of the angelic world, He none-the-less kept that eternal purpose a secret until the time was right to reveal it. It was finally through the apostle Paul that He revealed that secret. Another Pauline passage that touches on that purpose is Titus 1:2-3 saying "In hope of eternal life, which God, that cannot lie, promised before the world began; But hath in due times manifested his word through preaching, which is committed unto me according to the commandment of God our Saviour…"

Contrary to how it might appear to the casual reader of the Bible, man was not an after-thought by God. He not only had man in mind before he created the universe to begin with, but He also had His eternal purpose in mind that man would eventually be the eternal custodians of His creation as He states in Hebrews 2:7-9 "[5] For unto the angels hath he not put in subjection the world to come, whereof we speak. [6] But one in a certain place testified, saying, What is man, that thou art mindful of him? or the son of man, that thou visitest him? [7] Thou madest him a little lower than the angels; thou crownedst him with glory and honour, and didst set him over the works of thy hands: [8] Thou hast put all things in subjection under his feet. For in that he put all in subjection under him, he left nothing *that is* not put under him. But now we see not yet all things put under him."

Our God could do that (foresee the rebellion in the angelic world and the fall of man) because, unlike men or angels, He exists out side of His creation and can see and declare the end from the beginning (Isaiah 46:10). The fall in the angelic world did not catch God by surprise nor did the man's fall to the tempting of Satan. God planned for His ultimate victory. See in appendix 26 how the course of events in the redemption of man from history to prophecy layout. See also the book "You and Your Creator" by this same author for a study of how God's eternal purpose for His creation lays out in time and prophecy.

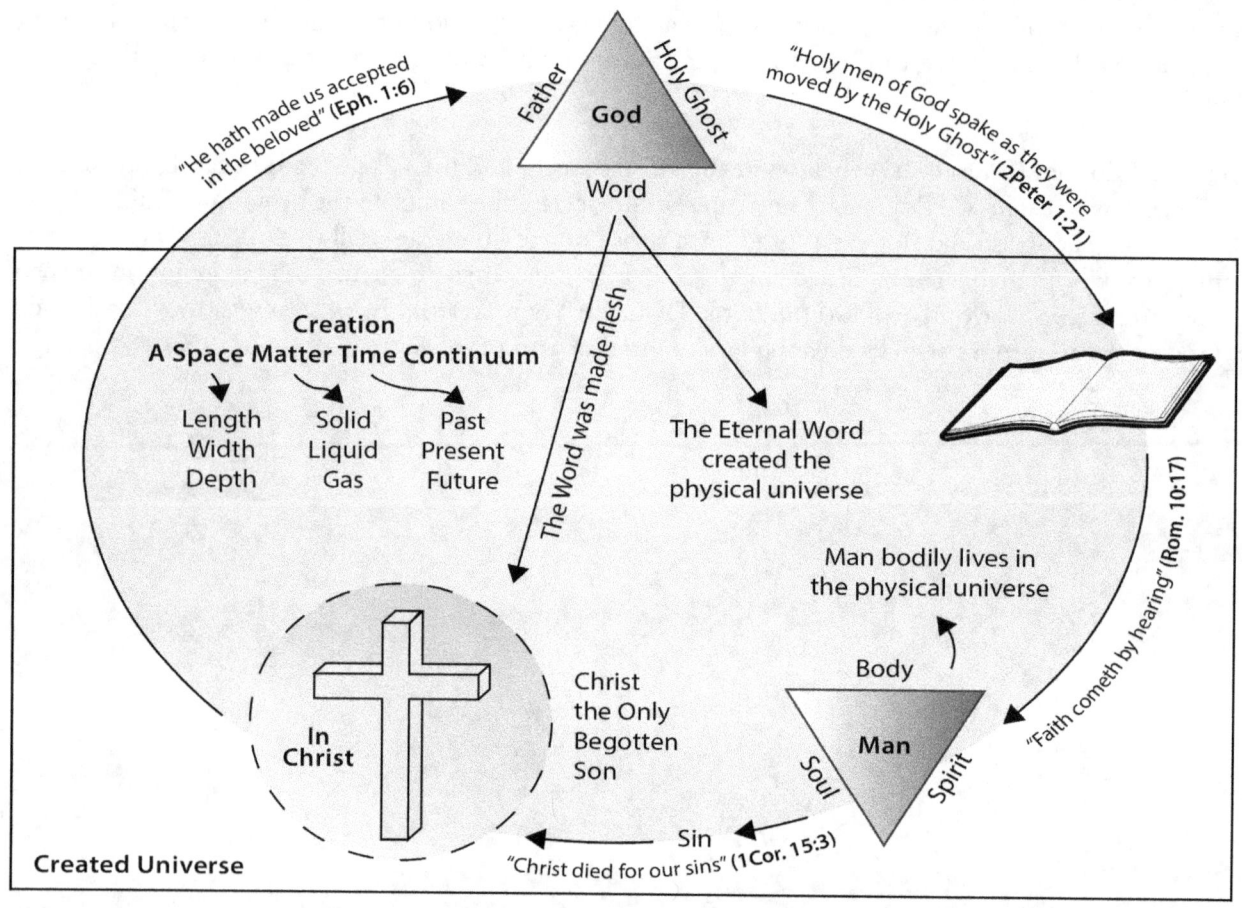

**Figure 1: Man – Created in the Image of His Creator**

# Genesis Annotations Chapter 1

## In the Beginning

The Bible text is presented in the body of this study with letters written in superscript denoting annotations on the specific words and concepts reference in the annotations which follow the Bible text. This study has a letter designation for each annotation. The letter designation refers to the letter superscript in the Bible text that corresponds to the annotation designated by the same letter for the first twelve chapters of Genesis. For example, the superscript [a] corresponds to the annotation designated [a] and so on throughout the study.

## Genesis 1:1-2 (KJV)

[1] In the beginning[a] God created[b] the heaven[c] and the earth. [2] And the earth was without[d] form, and void[e]; and darkness[h] *was* upon the face[i] of the deep[i]. And the Spirit[i] of God moved upon the face of the waters.

[a] **"In the beginning..."** This speaks of the beginning of time. The physical universe consists of a space – matter – time continuum. These three elements (a trinity of the physical creation) were created by God in what is here called, "the beginning." That is, the beginning of the space – matter – time makeup of the physical environment of which we in our physical body are a part.

- The gospel of John speaks of someone that was there already "in the beginning." John 1:1-10 "In the beginning was the Word and the Word was with God, and the Word was God. The same was in the beginning with God. All things were made by him; and without him was not anything made that was made…He was in the world, and the world was made by him, and the world knew him not…And the Word was made flesh, and dwelt among us (and we beheld of his glory, the glory as of the only begotten of the Father) full of grace and truth."
- Hebrews 1:10 "And, Thou, Lord, in the beginning hast laid the foundation of the earth; and the heavens are the works of thine hands." Compare this with verse 8 of Hebrews Chapter 1.
- 1 John 1:1 "That which was from the beginning, which we have heard, which we have seen with our eyes, which we have looked upon, and our hands have handled, of the Word of life;"

These all speak of Jesus Christ being the creator and the eternal God in that He created everything and anything that was created and that He was there from eternity past before the material universe existed. A further statement in Scripture of this creative act of God is found in Hebrews 11:3 "Through faith we understand that the worlds were formed by the Word of God, so that things which are seen were not made of things that do appear." In other words, the physical universe was created "out of nothing" – "ex nihilo" by the one who is called "the Word of God."

[b] **"God Created..."** – the word "created" is from the Hebrew "bara" This word is used only of God. Only He can truly create anything. Man can "make" and "form" things of previously existing things. At some point in the past, God created the heaven and the earth. At that point, time was created as well. Note the following regarding things that were created and things that God made out of what He had already created.

> Things God created ("bara"):
> Heaven, Genesis. 1:1
> Earth, Genesis 1:1
> Great Whales Gen 1:21 (the only creature named besides man)

Created Man (male and female), Genesis 1:27

Things God made ("basah") – out of previously existing material:
The firmament, Genesis 1:7
The Sun, Moon, Genesis 1:16
The beasts of the earth, Genesis 1:25
Trees, Genesis 2:9

**[c] "God Created the Heaven and the Earth"**… It does not say that God created the whole universe but it specifically refers to "the heaven and the earth." It is not until we come to Paul's epistles that we find a clue to God's reason for this two part reference. In Colossians 1:16 ff, we see that God had a purpose for each. We see in scripture that both are in the hands of the same usurper (2Cor. 4:4, Eph. 3:2). We see also that Christ will reconcile both to Himself; but the elect agency through which He will do so is different for each. The nation of Israel is God's means for the reclaiming of the earth while "the Church which is His body" is the means (His elect agency) whereby He reconciles the heavens back to Himself.

**[d] "And the earth was without form and void…"** Now in verse 2, God focuses on the earth. He does not focus to the heavens again until He reveals the mystery through Paul – the apostle of the Gentiles (2Cor. 5:1; Acts 26:19; 1 Cor. 15:48; Eph. 1:3, 10, 20; 2:6; 3:10; 2 Tim. 4:18; Phil. 2:10; 3:20; Col. 1:5, 16.). From Genesis 1:2 through to the middle of the Book of Acts, God deals with the earth and His earthly people Israel. From Genesis 11 onward until Paul comes on the scene, God's only dealings with Gentile nations is limited to how they are involved with Israel.

**[e] "And the earth was without form and void."** The word that is here translated "was" is translated "become or became" in Genesis 2:7; 18:18; and 19:26. Note that every verse in Genesis 1 begins with "and." The word "and" occurs 148 times in the first two chapters. They separate 102 separate acts of God. Each of them is an "and" of addition, not an "and" of further explanation. The event of verse one happened. Then, as a separate event, the event of verse two happened. There is a gap of time between the two events (i.e. between verses 1 and 2). Apparently, the angelic creation (the creation of the heavenly hosts – the angels, the cherubim, and the seraphim) was a part of the original creation of verse 1. The fall of Satan also apparently took place between verse 1 and verse 2 of Genesis 1. This would include all of the events of Isaiah 14:12-14 and Ezekiel 28:12-18. Job 38:4-7 indicates that the angels were present when God "laid the foundations of the earth…and laid the measures thereof…" Genesis 1:2 says the earth was without form and void (tohu and bohu). But Isaiah 45:18 says, "…he created it not in vain, [tohu] he created it to be inhabited." Therefore, we understand that the earth was where God intended to have His throne and probably did have it here between Genesis 1:1 and 1:2. God's throne today is in the third heaven (2Cor. 12:2). If there is a third heaven, there is then also a first and a second. The first would be the atmosphere of the earth. The second would then be the stellar universe. God had apparently moved His throne outside the stellar universe when sin entered it with the fall of Satan and the angels that followed him in rebellion.

**[f] Who inhabited the original earth? Who lived in the gap?** Ezekiel 28:13 indicates that Lucifer was in Eden before he fell and in fact was likely created there. Verse 14 of Ezekiel 28 indicates that he was "the anointed cherub that covereth." Wherever cherubim are seen in the Bible, they are somehow connected with the throne of God or with God's presence as in the mercy seat. God's purpose in the creation of Lucifer was apparently that he be the cherub that covered the throne. He was created very beautiful with musical instruments built into his very being (Ezek. 28:13). This would indicate that God Himself had His throne on the earth between Genesis 1:1 and 1:2. The "holy mountain of God" in Ezekiel 28:14 could very well be Mount Zion. Ezekiel 28:18 speaks of the fact that Lucifer "…hast defiled [his] sanctuaries by the multitude of [his] iniquities…" Lucifer's sanctuary would be the earth at that time. This accounts for the fact that the earth became without form and void in Genesis 1:2. God had moved His throne off the earth to the third heaven (2Cor. 12:2). Note that the first step in Lucifer's five point plan was "I will ascend into heaven..." (Isaiah. 14:13) That means his habitation was not in heaven. Heaven (which would have been at that time the stellar universe) was the habitation of angels. Cherubim are not angels and Lucifer being the anointed cherub is not and never was an angel.

The question is often raised in objection to the view that there was a gap between Genesis 1:1 and 1:2 by referring to Exodus 20:11 which states, "For [in] six days the LORD made heaven and earth, the sea, and all that in them [is], and rested the seventh day: wherefore the LORD blessed the Sabbath, and hallowed it." However, the key to understanding this might be in the distinction between "created" (bara – to create something out of nothing) and "made" (basah – to make using existing materials). Exodus 20:11 is referring to the six day re-creation of the earth to make ("basah") it a suitable habitation for man. When He did that, He did it using material that had been created in the beginning (that is in Genesis 1:1).

**[g] Without form and Void:** The term, "without form and void" is "tohu" and "bohu." We saw that it was not created that way (Isa. 45:18); but became that way by a cataclysmic judgment apparently in response to or in association with the fall of Lucifer. We find an interesting prophecy on the then eminent desolation of Judah by the captivity in Jeremiah 4:23-27. The LORD prophesied of the land being laid waste by recounting a desolation that happened to the earth much earlier in its history as a result of the rebellion of its earlier inhabitants. "I beheld the earth, and lo, it was without form and void; and the heavens, and they had no light. I beheld the mountains, and lo, they trembled, and all the hills moved lightly. I beheld the mountains, and all the birds of the heavens were fled. I beheld, and lo, the fruitful place was a wilderness, and all the cities thereof were broken down at the presence of the LORD, and by his fierce anger. For thus hath the LORD said, The whole land shall be desolate; yet will I not make a full end." The reference to "birds of the heavens" is interesting in Jeremiah 4:23. The birds that we know are only in the first heaven (the atmosphere of the earth). These birds referred to in Jeremiah 4:23 are the "birds of the heavens" (plural). This could well be a reference to "every unclean and hateful bird" of Revelation 18:2. There are winged creatures that occupy the "heavens" today. The cherubim in Ezekiel 1:6 and 10:20 each have four wings and four faces. We see references to cherubim and their wings in Exodus 25:20 and 37:9. The cherubim caricatured in 1Kings 6:24 each had two wings. The cherubim in Ezekiel 1:6 and 10:23 each have four wings and each four faces. The Seraphim in Isaiah 6:2 each had 6 wings. Two were used to cover their faces, two covered their feet and two were used to fly. We note that both cherubim and seraphim when found in Scripture are found in association with the throne of God.

**[h] The term "darkness" in verse 2** ties into a Pauline passage that speaks of that same darkness (2Cor. 4:6). It is interesting to note that Paul, of all Bible writers, is the one who takes us back to before the world began (Titus 1:2) to a promise that God made to Himself. The wisdom of God contained in the mystery was "ordained before the world unto our glory." (1Cor. 2:7; compare this with "from the foundation of the world" in Matthew 25: 34) We see in Ephesians that "...he hath chosen us in him before the foundation of the world..." (Eph. 1:4) In 2Timothy 1:9 we see that God "...hath saved us, and called us with an holy calling, not according to our works, but according to his own purpose and grace, which was given us in Christ Jesus before the world began, but is now made manifest..." This purpose has to do with the fact that there was sin and rebellion in the heavenly places and with the fact that the "heavens are not clean in his sight." (Job 15:15) The part of creation which was created by Christ and for Christ in the heavens (Col. 1:16) is in the hand of a usurper today (Eph. 2:2). But God is today calling out a body of believers that will one day live and reign in the heavens (2Cor. 5:1 cf. 2Tim. 2:10) for the honor and glory of Christ. The darkness then is the darkness of sin and ruin. This is the darkness that Paul speaks of being dispelling by the light in 2Corinthians 4:6.

**[i] "...Darkness was upon the face of the deep.** And the Spirit of God moved upon the face of the waters."
Two questions come to mind as we read this. One is: What is the "Deep"? The second would have to do with this reference to water and it being separation into two bodies in verse 7 – one being above the firmament and the other being below the firmament.

The deep is obviously a reference to the earth being covered by water. But as we study this passage in light of other Bible passages, we find that there is more to it than that there was simply a body of water covering the entire earth. "The face" of the deep is simply the surface of it. The passages in Genesis 7 on the face of the waters of the flood make this clear:

Genesis 7:18 – The ark went upon the face of the waters

Genesis 7:23 – Every living substance destroyed from the face of the ground.

Psalm 104:1-9 says a lot about the waters, the deep and the light:

¹ Bless the LORD, O my soul. O LORD my God, thou art very great; thou art clothed with honour and majesty. ² Who coverest *thyself* with light as *with* a garment: who stretchest out the heavens like a curtain: ³ Who layeth the beams of his chambers in the waters: who maketh the clouds his chariot: who walketh upon the wings of the wind: ⁴ Who maketh his angels spirits; his ministers a flaming fire: ⁵ *Who* laid the foundations of the earth, *that* it should not be removed for ever. ⁶ Thou coveredst it with the deep as *with* a garment: the waters stood above the mountains. ⁷ At thy rebuke they fled; at the voice of thy thunder they hasted away. ⁸ They go up by the mountains; they go down by the valleys unto the place which thou hast founded for them. ⁹ Thou hast set a bound that they may not pass over; that they turn not again to cover the earth.

Verses 6 and 7 of the 104th Psalm speak of God covering the earth with the deep as with a garment. This deep in not the oceans for this deep covered the mountains. The waters referred to in this passage fled (i.e. fled from off the earth) at God's voice (vs.7).

There was a cataclysmic judgment that destroyed the original earth by water. The Book of Job offers some insight into the original earth and the re-creation of it to make a suitable place for man. The Book of Job was written before Moses wrote Genesis.

Let's consider Job 38:4-11

"Where wast thou when I laid the foundations of the earth? declare, if thou hast understanding.
⁵Who hath laid the measures thereof, if thou knowest? or who hath stretched the line upon it?
⁶Whereupon are the foundations thereof fastened? or who laid the corner stone thereof;
⁷When the morning stars sang together, and all the sons of God shouted for joy?
⁸Or who shut up the sea with doors, when it brake forth, as if it had issued out of the womb?
⁹When I made the cloud the garment thereof, and thick darkness a swaddling band for it,
¹⁰And brake up for it my decreed place, and set bars and doors,
¹¹And said, Hitherto shalt thou come, but no further: and here shall thy proud waves be stayed?"

Verse 8 speaks of a time when a sea enveloped the earth.
Verse 9 speaks of the deep thus formed being enveloped by darkness.
Verse 7 tells us that the sons of God and the morning stars (i.e. angels and angelic creatures) were there when that happened. This is a reference to something that happened between Gen.1:1 and Gen. 1:2.

Job 38:30 "The waters are hid as with a stone, and the face of the deep is frozen." The face of the deep is today frozen. Is this simply a reference to the fact that some lakes freeze (as some see this passage) or does it contain more information than that? We might consider Job 37:18 where we find that the sky is strong as a molten looking glass: "Hast thou with him spread out the sky, which is strong, and as a molten looking glass?" A looking glass has two sides to it. The question is then if the sky is as a looking glass, what is above it and what is below it? Well we know what is below it – The stellar universe including our galaxy and our solar system. Above that would then be outside of the universe. This would be what the apostle Paul called the third heaven in 2Corinthians 12:2. In Revelation 4:4-6 we gain some insight into this: "And before the throne there was a sea of glass like unto crystal: and in the midst of the throne and round about the throne, were four beasts full of eyes before and behind." Here the apostle John is given a vision of the throne of God and describes the area before it a sea of glass. This would then be the face of the deep that is frozen. This is the floor of the third heaven and the underside of which we see as sky.

Job 26:7 -13 also sheds light on the subject of the deep and the separation of the third heaven from the universe that we can look at from the earth.

> 7 He stretcheth out the north over the empty place, *and* hangeth the earth upon nothing. 8 He bindeth up the waters in his thick clouds; and the cloud is not rent under them. 9 He holdeth back the face of his throne, *and* spreadeth his cloud upon it. 10 He hath compassed the waters with bounds, until the day and night come to an end. 11 The pillars of heaven tremble and are astonished at his reproof. 12 He divideth the sea with his power, and by his understanding he smiteth through the proud. 13 By his spirit he hath garnished the heavens; his hand hath formed the crooked serpent. 14 Lo, these *are* parts of his ways: but how little a portion is heard of him? but the thunder of his power who can understand? Job 26:6-14

Note: Verse 7 of Job 26 speaks of God stretching the north over the empty place. This would be a reference to the sides of the north where the angels meet (Isa. 14:13 - 14). This is likely the "Mount of the Congregation" where the sons of God (angels) met in Job 1:6 & 2:1 to present themselves before the Lord.

Note also that "...He bindeth up the waters in His thick cloud... He holdeth back the face of His throne." He has hidden His throne from view "...until the day and night come to an end."(vs 10) This day and night coming to an end will happen when in the New Jerusalem "...there shall be no night there..." (Rev. 21:25).

## The Deep

The Deep can be a reference to the oceans and seas but can also refer to the waters above the heavens. The references (below) to the deep are listed with regard to the heavens or the earth:

   The deep with regard to the heavens:
   - Genesis 7:11 "...the same day were all the fountains of the great deep broken up and the windows of heaven were opened."
   - Genesis 8:2 "The fountains also of the deep and the windows of heaven were stopped, and the rain from heaven was restrained."
   - Deuteronomy 33:13 "And of Joseph he said, Blessed of the LORD he his lord, for the precious things of heaven, for the dew, and for the deep that coucheth beneath..."
   - Job 38:30 "The waters are hid as with a stone, and the face of the deep is frozen.
   - Proverbs 8:28 "When he established the clouds above: when he strengthened the fountains of the deep: when he gave to the sea his decree, that the waters should not pass his commandment...I [wisdom] was there..."

   The deep with regard to the earth: Genesis 1:2 "Darkness was upon the face of the deep."
   - Job 41:31 "He maketh the deep to boil like a pot: he maketh the sea like a pot of ointment. He maketh a path to shine after him; one would think the deep to be hoary."
   - Psalm 104:6 "Thou coveredst it with the deep as with a garment: the waters stood above the mountains. At thy rebuke they fled; at the voice of thy thunder they hasted away."
   - Psalm 107:24 "These see the works of the LORD, and his wonders in the deep. For he commandeth, and raiseth the stormy wind, which lifteth up the waves thereof."
   - Isaiah 44:27 "That saith to the deep, be dry, and I will dry up thy rivers..."

- Isaiah 51:10 "Art thou not it which hath dried the sea, the waters of the great deep; that hath made the depths of the sea a way for the ransomed to pass over?"
- Isaiah. 63:13 "He [Moses] led them through the deep…"

## Dividing the waters from the waters (Gen. 1:7)

Genesis 1:7-8 "And God made the firmament, and divided the waters which were under the firmament from the waters which were above the firmament and it was so. And God called the firmament Heaven." The firmament is where the sun, moon and stars exist.

Genesis 1:9-10 "And God said, Let the waters under the heaven [the firmament] be gathered together unto one place, and let the dry land appear; and it was so. And God called the dry land Earth; and the gathering together of the waters called he Seas…" This was done in the preparation of the earth to receive life.

Genesis 1:14 says "Let there be lights in the firmament of the heaven to divide the day from the night." The purpose for the sun moon and stars was to give light on the earth and to demark day from night.

**[j] "And the Spirit of God moved upon the face of the waters."** Here we find the Spirit of God (the Holy Spirit) begin to remake the ruined earth so as to prepare it as a suitable habitation for man. Hebrews 9:14 refers to the Holy Spirit as the eternal Spirit.

## Genesis 1:3-5 (KJV) Day One of Creation Week

[3] And God said[k], Let there be light[l] and there was light. [4] And God saw the light, that *it was* good[m] and God divided[n] the light from the darkness. [5] And God called the light Day, and the darkness he called Night. And the evening and the morning were the first day.

**[k] "…And God said…"** – The power of God's command is here demonstrated. "Let" (God calleth those things that be not as though they were…Rom. 4:17). This phrase "…and God said" is used ten times in Genesis 1. The term "Let" is used eight times. Then "Let us make" is used in verse 26 "Let us make man in our image." The power of God's Word is infinite. The true believer has the faith that saves because of his faith in that same power. "[3] Through faith we understand that the worlds were framed by the word of God, so that things which are seen were not made of things which do appear." (Heb. 11:3) "Let" implies permission for something that has been carefully planned to happen. Creation had been planned in the mind of God. All that was needed now was for the time to be right to let it happen and God acted to implement His plan.

## "Let" – speaks of an event being given permission to happen:

This word "Let" speaks of a process. Let's consider what happened on the third day when the earth was caused to bring forth the biosphere: "[11] And God said, Let the earth bring forth grass, the herb yielding seed, *and* the fruit tree yielding fruit after his kind, whose seed *is* in itself, upon the earth: and it was so. [12] And the earth brought forth grass, *and* herb yielding seed after his kind, and the tree yielding fruit, whose seed *was* in itself, after his kind: and God saw that *it was* good. [13] And the evening and the morning were the third day." (Genesis 1:11-13) This is a reference to a supernatural process in which the natural process is overridden by God's creative ability. The mineral content of the earth was supernaturally incorporated into the living biomass that we see it on earth today. To visualize this we need to think of time lapse photography of (for example) an apple seed in soil that would germinate, sprout, grow to full development, develop foliage, bloom, and bear fruit all in one day of 24 hours in duration. Then picture this same scenario would have been repeated many time concurrently in that same day of creation week as the entire biomass of the earth was brought forth to fully functional maturity.

Those same carefully planned processes would have been duplicated on each day of creation week. If we would have been there we would have had an experience of witnessing in the formation of animal life (the zoo sphere) similar to what Ezekiel had in Chapter 37 of Ezekiel in the vision of the valley of the dry bones. Ezekiel saw as the dry bones of the dead of Israel first had sinew come upon them, and then God brought flesh upon them, and then the skin, and then God put breath in them, and the dry bones lived, and stood upright and lived. Thus we can understand how every detail of what was to happen on each of those days was planed meticulously and executed to perfection. On the second day of creation week, one would watch the portion of the water that was removed from the earth being taken up in what would be similar to what happed in Genesis 8:13 when the water that produced the flood was removed from off of the earth.

Note the usage of the term "Let." It was not that God simply said "Let it be" and it was. He used a process to bring it about. Adam would have witnessed that on the sixth day of creation week when he witnessed God create the animals as we will see it in Genesis 2:16. All of creation week involved what looked like natural processes but in a supernatural manner in that it was done by a supernatural acceleration of those natural processes. When the six days of that week was over, the earth was filled with the fully mature creation that exists to this day. The earth was filled and ready to be ruled by this new race of free moral agent – man who was charged with the responsibility to rule in it for the honor and glory of his creator.

Table 1:" And God Said…"

| Day | "Let" | Genesis | Quote |
|---|---|---|---|
| 1 | 1 | 1:3 | "Let there be light" |
| 2 | 2 | 1:6 | "Let there be a firmament in the midst of the waters…" |
| 3 | 3 | 1:9 | "Let the waters under the heaven be gathered together in one place…" |
| 3 | 4 | 1:11 | "Let the earth bring forth grass, the herb yielding seed…" |
| 4 | 5 | 1:14 | "Let there be lights in the firmament of heaven to divide the day from…" |
| 5 | 6 | 1:20 | "Let the waters bring forth abundantly the moving creature that has life…" |
| 6 | 7 | 1:24 | "Let the earth bring forth the living creature after his kind…" |
| 6 | 8 | 1:26 | "Let us make man in our image, after our likeness: and let them have dominion…" |
| 6 | | 1:28 | "Be fruitful and multiply and replenish the earth…" |
| 6 | | 1:29 | "I have given you every herb…" |

Those same carefully planned processes would have been duplicated on each day of creation week. If we would have been there we would have had an experience of witnessing in the formation of animal life (the zoo sphere) similar to what Ezekiel had in Chapter 37 of Ezekiel in the vision of the valley of the dry bones. Ezekiel saw as the dry bones of the dead of Israel first had sinew come upon them, and then God brought flesh upon them, and then the skin, and then God put breath in them, and the dry bones lived, and stood upright and lived. Thus we can understand how every detail of what was to happen on each of those days was planed meticulously and executed to perfection. On the second day of creation week, one would watch the portion of the water that was removed from the earth being taken up in what would be similar to what happed in Genesis 8:13 when the water that produced the flood was removed from off of the earth.

## The Evening and the Morning

Table 2:: "And the evening and the morning" used 6 times.

| Sphere | Genesis | How each day closes: |
|---|---|---|
| Atmosphere | 1:5 | 1 closes with God commanding the light to shine out of darkness |
| Hydrosphere | 1:8 | 2 closes with God dividing the waters to make the firmament – the open space |

| Lithosphere | 1:13 | 3 closes with the appearance of dry land and the vegetative life. |
| Stratosphere | 1:19 | 4 closes with God making the sun and the moon and causing stars to appear |
| Biosphere | 1:23 | 5 closes with the creation of all the fish and fowl life in the earth |
| Zoosphere | 1:31 | 6 closes with the creation of land animals and man |
| | | The seventh day, the Sabbath, has no evening because it speaks of eternal, unending rest (Heb. 4:5). |

## [l] The source of the light on the first day:

This light is not from the sun. "Thou hast prepared the light <u>and</u> the sun" (Psalm 74:16). The sun was created on the fourth day. This is the light that God commanded to shine to dispel the darkness of a ruined creation (2 Cor. 4:6). This is the power of the Word of God to give light to the darkness of unbelief. Just as His command brought forth light to shine in the darkness, so His Word does in the heart upon belief.

"Light is sown for the righteous" (Psalm 97:11).
"Who coverest thyself with light as with a garment" (Psalm 104:2).
"I form the light, and create darkness" (Isa. 45:7).

The real light is not that of the sun but of the Son.

"The sun shall be no more thy light by day; neither for brightness shall the moon give light unto thee…And I saw no temple therein: for the Lord God Almighty and the Lamb are the temple of it. And the city had no need of the sun, neither of the moon, to shine in it for the glory of God did light it and the Lamb is the light thereof." (Revelation 21:22-23).

Paul says of Christ that He is "dwelling in the light which no man can approach unto…" (1Tim. 6:19). Believers today are "children of light" (Eph. 5:8). The significance of the light here in Genesis 1:4 is that God shined a spotlight to light the earth to call attention to the work that He will do in redeeming it from the corruption of sin. Angels were watching this work of God (Job 38:7) - thus the light is there to draw attention to what God is doing.

## [m] "…it was good…"

God saw the light that it was "good." Note: this is not an absolute statement. See Genesis 2:18. The term is used 7 times in this chapter: It is the statement of a craftsman who sits back after completing a masterpiece and looking back at it with satisfaction that it is the way he wanted it.

Table 3: "It is Good…"

| Gen. 1:4 | With respect to the light (but does not call the darkness good) |
|---|---|
| Gen. 1:10 | With respect to the dividing of sea from land (does not call the heavens good (cf Job 15:15) |
| Gen. 1:12 | With respect to the vegetable creation |
| Gen. 1:18 | With respect to the making of the sun and moon |
| Gen. 1:21 | With respect to the making of birds and fish |
| Gen. 1:25 | With respect to the making of land animals |
| Gen. 1:31 | With respect to the creation of man |

## [n] Dividing the light from darkness

God divided light from darkness, There is a parallel of the contrasts of light with darkens and of wisdom with folly in Scripture (e.g. Eccl. 2:13; 11:7-8; Matt. 4:16; Luke 1:79; 11:34; John 1:4-5; 3:19; 12:46; Act 26:18; Rom. 13:12; 2 Cor. 6:14; etc.). God's work on the first day: He "prepared" the light and "made" the darkness. Believers are "children of light and children of the day: we are not of the night, nor of darkness." (1Thess. 5:5) In a spiritual sense, we have light to see in the darkness and to dispel the darkness. Believers are to be the light of the world (Eph. 5:13).

We might consider the following verses on the contrast between light and darkness:

Genesis 8:22 "while earth remaineth, seed time and harvest…and day and night…shall not cease"
Psalm 19:2 "Day unto day offereth speech, and night unto night showeth knowledge."
Psalm 74:16 "The day is thine, the night is thine: thou has prepared the light and the sun."
Psalm 104:20 "Thou makest the darkness and it is night:…"
Isaiah 45:7 "I form the light and create darkness"
1Corinthians 3:13 "…the day shall declare it…" -- the Day of Judgment contrasted with the "darkness of this day"

## Genesis 1:6-8 (KJV) The Second Day

6 And God said, Let there be a firmament[o] in the midst of the waters, and let it divide the waters from the waters. 7 And God made the firmament, and divided the waters which *were* under the firmament from the waters which *were* above the firmament: and it was so. 8 And God called the firmament[p] Heaven. And the evening and the morning were the second day.

## [o] The firmament of Genesis 1:6:

God's work on the second day was to "make" a firmament in the midst of the waters. The word "firmament" means simply "open space." God divided the waters that comprised "the deep" that covered the earth. Considering passages as Job 37:18 and 38:30, it appears that "the deep" as being still there but God made an open space (i.e. a firmament) in the midst of it. Our universe exists in this open space

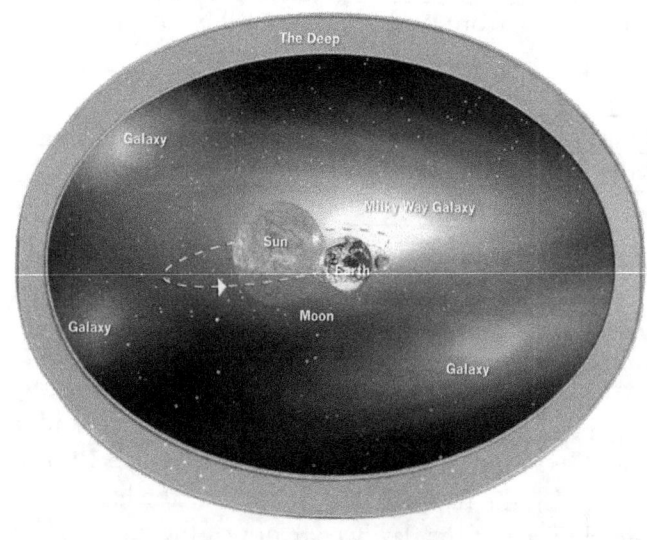

**Figure 2:  The Deep**

Many creation scientists talk about a vapor canopy that once covered the earth.  It could be that there was such a canopy of water vapor. That would partially explain the long life span of people before the flood.  However, if there was, it would have been a part of the waters that were below the firmament because the firmament encompassed the sun, the moon and the stars. A vapor canopy would do a number of things beneficial to the life span of people on earth:

1. Filter out ultraviolet radiation as well as other cosmic radiation.  This would reduce the effect of somatic mutation – a primary cause of aging.

2. The atmospheric pressure would be greater.  Therefore, the body would probably function better. It could be that this canopy will be reestablished in the new earth.

There would be more of a green house effect on the earth to produce a fairly uniform climate.  Those would include:

> No extremes of heat and cold.
> Global green house producing lush vegetation
> Incoming solar radiation retained and dispersed
> Uniform temperature
> No movement of large air masses
> No violent weather
> No hydrologic cycle
> No dust in upper atmosphere-no precipitation
> Vapor cycle consisting of alternating evaporation and condensation
> Lush vegetation the world over
> UV and cosmic radiation filtered
> Greater atmospheric pressure

It should be noted however that the firmament (the open space) that was formed on the second day accommodates not only the moon but the sun also (Gen. 1:14-15).  Therefore, if this refers to the vapor canopy that collapsed to produce the flood of Genesis 7, then it would be beyond the sun.  Therefore, the UV filtering effect would not be a valid argument.  This firmament divided the waters which were above the firmament from the waters which were under the firmament (i.e. on earth). The waters which were above the firmament are beyond our solar system (verses 14-15), and likely beyond our galaxy and probably beyond the stars as well.

**[p] #16 The firmament is called heaven:** God called the firmament "Heaven." This is the third thing named by God in the Bible. God-gave names to a number of things:

Table 4: the Daily Log of Creation Week

| Day | Verse | Named |
|---|---|---|
| 1 | Gen. 1:5 | God called the light "Day" |
| 2 | Gen. 1:5 | The darkness he called "Night" |
| 3 | Gen. 1:8 | God called the firmament "Heaven" |
| 4 | Gen. 1:10 | God called the dry land "Earth" |
| 5 | Gen. 1:10 | The gathering together of the waters called He "Seas" |
| 6 | Gen 2:16 | Adam "called" the creation by name |
| 7 | Gen 3:20 | Adam "called" his wife "Eve" |
| | | God called their name Adam |

This heaven in verse 8 however is not the heaven of verse 1. That heaven in verse 1was created in the beginning. We will find as we go on in this study that there are three heavens referenced in the Bible (2Cor. 12:2).

1st Heaven – The atmosphere of the earth (Gen. 1:20)
2nd Heaven – The stellar universe (Gen. 15:5; 22:17; Exodus 32:13)
3rd Heaven – The throne of God (2Cor. 12:6; Rev. 4:6)

## Genesis 1:9-13 (KJV) The Third Day

⁹ And God said, Let the waters under the heaven be gathered[q] together unto one place, and let the dry *land* appear: and it was so. ¹⁰ And God called the dry *land* Earth; and the gathering together of the waters called he Seas: and God saw that *it was* good. ¹¹ And God said, Let the earth bring forth[r] grass, the herb yielding seed, *and* the fruit tree yielding fruit after[s] his kind, whose seed *is* in itself, upon the earth: and it was so. ¹² And the earth brought forth grass, *and* herb yielding seed after his kind, and the tree yielding fruit, whose seed *was* in itself, after his kind: and God saw that *it was* good. ¹³ And the evening and the morning were the third day.

**[q] Dividing the waters on the third day:**

Job 26:7-13

"⁷He stretcheth out the north over the empty place, and hangeth the earth upon nothing.
⁸He bindeth up the waters in his thick clouds; and the cloud is not rent under them.
⁹He holdeth back the face of his throne, and spreadeth his cloud upon it.
¹⁰He hath compassed the waters with bounds, until the day and night come to an end.
¹¹The pillars of heaven tremble and are astonished at his reproof.
¹²He divideth the sea with his power, and by his understanding he smiteth through the proud.
¹³By his spirit he hath garnished the heavens; his hand hath formed the crooked serpent.

God's first act on the third day was to focus on the waters under the heaven (i.e. the waters on the earth) to be gathered together in one place. Likely, before the flood, there was only one body of water or one sea on earth. The reference to seas (plural) is perhaps a reference to the "sea" above the firmament and the "sea" under the heaven. The "sea" referred to in Job 38:8-11 could be the sea above the firmament of which "the cloud" (verse 9) is the garment thereof (cf. Psalm 24:2). Psalm 95:5; 104:34-39; 33:7; Jeremiah 5:22; etc. speak of the sea on earth. 2Peter 3:5 is talking about the flood of Noah's day but might have an application to the separation of the waters and the forming of dry land of Genesis 1 as well.

**[r] Vegetation created on the third day:**
God, as it were, speaks a command to the earth to bring forth vegetative matter. God here takes the minerals of the earth and fashions all of the various living plants. Man can manipulate life, but only God can produce life from inorganic substances. Also, He designed a system whereby the living vegetation can reproduce itself by means of "seed" that was "in itself." However, the reproduction was only "after his kind."

**[s] Three classes of vegetative** matter are viewed here: Grasses, herbs, and trees.

## Genesis 1:14-19 The Fourth Day

[14] And God said, Let there be lights in the firmament of the heaven to divide the day from the night; and let them be for[z] signs, and for seasons, and for days, and years: [15]And let them be for lights in the firmament of the heaven to give light upon the earth: and it was so. [16] And God made two[t] great lights; the greater light to rule[u] the day, and the lesser[v] light to rule the night: he [w]made the stars[x] also. [17] And God set them in the firmament[y] of the heaven to give light upon the earth, [18] And to rule over the day and over the night, and to divide the light from the darkness: and God saw that it was good. [19]And the evening and the morning were the fourth day.

**[t]. Two different lights are referenced in Genesis Chapter 1:**
The word for lights in verse 14 is different than the word used in verse 3. The word in verse 14 indicates "light sources." The sun and the moon are light sources. The light of verse 3 was a special light that God used to draw attention to his work of creation week. Job 38:9 indicates that the angelic hosts were observing the events of creation week. The light of verse 3 was a spotlight for them to see what was happening. The lights of verses 14-17 were light sources that would continue to provide light after creation week. Psalm 74:16 explains this saying "he prepared the light <u>and</u> the sun." The ordinances of the sun and the moon, of day and night will last forever. (Gen. 8:22; Jer. 31:35)

**[u] The purpose for the sun, moon and stars:**


God's purpose for the sun, the moon, and the stars is here stated as being for:

- Signs – The lunar cycle of 30 days. Cf Psalm 81:3; Joel 2:10; 3:15. cf Revelation 6:12 and 8:12
- Seasons – Spring, Summer, Fall, Winter (The earth's elliptical orbit)
- Days – The rotation of the earth on its axis.
- Years – The 360 degrees of the compass ☐The revolution of the earth about the sun.

**[v] Sun and moon are in the firmament:**

The "firmament" apparently is beyond the sun for the sun is in "the firmament."
See note [o].

**[w] The sun and moon were "Made…"** – That is to say that they were made from existing material that was created in the beginning (Genesis 1:1).

God "made" the sun and the moon (see note [b]). They were made of material created "in the beginning." Neither the sun nor the moon were part of the original creation of Genesis 1:1 but were made on the fourth day. It is this author's opinion that the current sun is a placeholder for the new Jerusalem that the Revelation speaks of. See the author's book *A Study in the Revelation* (2023). If the universe is heliocentric, then the New Jerusalem will be the center of the New Heaven. Also, we can then understand how "the nations of the earth will walk in the light of the New Jerusalem" (Rev. 21:24). "And the city had no need of the sun, neither of the moon, to shine in it: for the glory of God did lighten it, and the Lamb is the light thereof." (Revelation 21:23)

**[x] "The stars also…"**

The stars are often associated with angels in Scripture. For example, the great red dragon on Revelation 12:3 & 4 having seven heads and ten horns and seven crowns on his heads (who we understand to be Satan) drew a third part of the stars from heaven with his tail. We understand this to mean that one third of the hosts of heaven (one third of the angels) followed Lucifer in rebellion. We can compare Genesis 1:16 & 17 with Psalm 136:8-9 and note the words that the translators added (words added by the translators are in italics in Genesis 1:16). It is apparent that the stars which were created in the beginning would rule the night along with the moon which was made on day 4 of creation week.

### Genesis 1:16-18 Compared with Psalm 136:7-8 Regarding The Fourth Day

16 And God made two great lights; the greater light to rule the day, and the lesser light to rule the night: *he made* the stars also. 17 And God set them in the firmament of the heaven to give light upon the earth, 18 And to rule over the day and over the night, and to divide the light from the darkness: and God saw that *it was* good. (Gen 1:16-18)

7 To him that made great lights: for his mercy *endureth* for ever: 8 The sun to rule by day: for his mercy *endureth* for ever: 9 The moon and stars to rule by night: for his mercy *endureth* for ever. (Psalms 136: 7&8)

The stars were created "in the beginning" as part of the heaven. The reference here in the italicized words (*he made)* is that the stars are made to rule the night along with the moon. Psalm 136:8-9. "The sun to rule the day… the moon and stars to rule the night."

**[y] The sun and the moon are in the firmament of heaven**
God set them (the sun and the moon) in the firmament of <u>the heaven</u>. Therefore, the firmament, the open space between the waters extends beyond our solar system and likely beyond the stars as well.

### Genesis 1:20-23 The Fifth Day

20 And God said, Let the waters bring forth abundantly the moving creature that hath[z] life, and fowl *that* may fly above the earth in the open firmament[aa] of heaven. 21 And God created[ac] great whales, and every living creature that moveth, which the waters brought forth abundantly, after their kind, and every winged fowl after his kind: and God saw that *it was* good. 22 And God blessed them, saying, Be[ad] fruitful, and multiply, and fill the waters in the seas, and let fowl multiply in the earth. 23 And the evening and the morning were the fifth day.

**[z] The stars tell a story**

The lights are for "signs."  A study of Psalm 19 indicates that the gospel (a primitive gospel) may have been presented in the stars up until the Bible was written.

Psalm 19:1-6: The stars communicate.
Psalm 19:7-14: The law [written word] communicates.
Genesis 37:9: Sun, moon, and eleven stars made obeisance to Joseph.
Psalm 147:4 and Isaiah 40:36: The stars are "numbered" and "named."
BUT!  When God withdrew these signs, man corrupted them.
Isaiah 47:13 speaks of the stars being a part of a perverted and corrupted system of worship. "…let now the astrologers, the stargazers, the monthly prognosticators stand up, and save thee from these things…

**[aa] The first life appears on the fifth day**

Life appears on the fifth day.  Christ appears in the start of the fifth millennium to give eternal life. (John 3:15; 5:39; 6:54; 6:68; 10:28; 12:25; 17:2; Rom. 5:21; 1 Tim. 1:16; etc.)

**[ab] The life in the air**

The word "fowl" means anything that flies (birds or insect).  The open firmament of heaven here is the first heaven—the earth's atmosphere.

**[ac] Life is created not made**

God "created" (bara – see note [b]) life.  Other things (aside from the original creation of the heaven and the earth) were "made" (basah) of previously existing material.

## Genesis 1:24-25 The Sixth Day

> [24] And God said, Let the earth bring forth the living creature after his[ad] kind, cattle, and creeping thing, and beast of the earth after his kind: and it was so. [25] And God made the beast[ae] of the earth after his kind, and cattle after their kind, and every thing that creepeth upon the earth after his kind: and God saw that *it was* good.

**[ad] A three fold commission**

The fowl and fish life has a three-fold commission:

- Be fruitful
- Multiply
- Fill the waters and air.

**[ae] Animal life appears on the sixth day along with man.**

Three groups of animals are listed:

- Cattle – domestic animals
- Creeping things – reptiles, worms, etc.
- Beasts – wild animals

## Genesis 1:26-28 The Sixth Day Continues

26 And God said, Let us make[ag] man in our[af] image, after our[ag] likeness: and let them have dominion[ai] over the fish of the sea, and over the fowl of the air, and over the cattle, and over all the earth, and over every creeping thing that creepeth upon the earth. 27 So God created[ag] man in his[ah] *own* image, in the image[ai] of God created[ai] he him; male[ak] and female created he them. 28 And God blessed[al] them, and God said unto them, Be[am] fruitful, and multiply, and replenish the earth, and subdue it: and have[ai] dominion over the fish of the sea, and over the fowl of the air, and over every living thing that moveth upon the earth.

### [af] Man is created in the image and likeness of God

The term "Let us make man in our image…" speaks of a plurality in the one God (that is in reference to what we call the trinity). Other occurrences of such an expression that has reference to the trinity include:

Psalm 2:7 "I will declare the decree: the LORD hath said unto me, Thou art my Son; this day have I begotten thee."Isaiah 48:15-16 "…I have spoken; yea, I have called him: I have brought him…"

Psalm 45:7 "Thou lovest righteousness, and hatest wickedness: therefore God, thy God, hath anointed thee with the oil of gladness above thy fellows." Psalms 45:7

Psalm 110:1 "The LORD said unto my Lord, Sit thou at my right hand, until I make thine enemies thy footstool."

Matthew. 11:27 " All things are delivered unto me of my Father: and no man knoweth the Son, but the Father; neither knoweth any man the Father, save the Son, and *he* to whomsoever the Son will reveal *him*."

John 8:42 "Jesus said unto them, If God were your Father, ye would love me: for I proceeded forth and came from God; neither came I of myself, but he sent me."

Genesis 3:22 "And the LORD God said, Behold, the man is become as one of us, to know good and evil: and now, lest he put forth his hand, and take also of the tree of life, and eat, and live for ever:…"

Genesis 11:7 "Go to, let us go down, and there confound their language, that they may not understand one another's speech."

### [ag] Man is both "made" and "created"

Note from verses 26 and 27 that man was both created and made. Man was both "made" (basah) of previously existing material and he was also "created" (bara) – having been brought into existence out of nothing (cf note

### [ah] The term "image of God" in Verse 27 speaks of ownership

Just as the coin that bore Caesar's image belongs to Caesar (Matt. 22:19; Luke20:24), so the creature that bares God's image belongs to God. Therefore, man belongs to God, his Creator.

### [ai] Man is both "Created" and he is also "Made." He was "made" with respect to the brain and the physical body.

- His body was "formed" as the bodies of animals were (Gen. 1:24; 2:7).
- He has the "breath of life" as do animals (Gen. 2:7; 7:22).

- He is a "living soul" as are animals (Gen. 1:24: 2:7).

He was "created" with respect to the spirit and the soul.

- But there is a difference between the spirit of man and the spirit of animals (Ecclesiastes 3:21) – the spirit of man goes upward but the spirit of beasts goes downward.
- Man's spirit has the capacity to know, to love and to worship God (John 4:23, 24)

There are three mentalities in man that correspond to the three part makeup of man:

1. His Spirit has what the Bible calls a <u>mind</u> that knows things and gathers knowledge – particularly things that pertain to God and how man can relate to his creator (Rom. 8:26)
2. His Soul has a mentality called in the Bible the <u>heart</u> that thinks, makes decisions in life (Rom. 10:10) and understands (Psa. 23:7).
3. His Body has a brain and the five senses that enable him to carry out the decisions made in the heart (the soul) in the real world through his physical body. It is in our physical bodies that we carry out the will of the soul and spirit. However, the mentality of the body is called "the flesh" in Scripture. It is in the physical body that the sin nature resides in the three part of the human makeup (Rom. 7:18).

## Man in the image of God

The <u>image of God</u> in the Bible is ultimately found in the person of Jesus Christ. He is "the image of God" (2 Cor. 4:4). He is "...the brightness of his [God's] glory and the express image of his person..." (Heb. 1:3). In 1Corinthians 11:7 we find it stated that man [male of the human race] is "...the image and glory of God: but the woman is the glory of the man." That is not to say that the woman is not in the image of God. She was created in the image of God in Adam but was taken out of Adam and is therefore said to be the glory of the man. Verse 27 makes it clear that both are in the image of God. "So God created <u>man</u> in his own image, in the image of God created he <u>him</u>, male and female created he <u>them</u>."

The image of God needs to be understood. In scripture we find that God is a trinity. Three persons called God yet there is but one God (Deut. 6:4).

- The Father – John 1:14; 6:23
- The Son – John 1:18; 3:35; Heb.1:8
- The Spirit – Acts 5:4

Compare these with 1 John 5:7 "...the Father, the Word, and the Holy Ghost: and these three are one."

We find also that Man is a trinity of body, soul, and spirit. (1 Thess. 5:23) We find that trinity of man in Genesis 2:7.

- Body – formed out of the dust of the ground
- Spirit – Breathed into his nostrils the breath of life
- Soul – Man became a living soul.

We can find also in Scripture that God has:

- A Body that could be seen (Exodus 24:10; Rev. 4:5)
- A Soul that is well pleased with Jesus the Son (Matt. 12:18)
- A Spirit which receives the worship of men (John 4:24)

After the fall, we find that Adam begot a son in His own likeness, after his image (Gen. 5:1-3). Yet, in Genesis 9:6, we find that man is still in the image of God though that image has been marred by sin. Paul says in Acts 17:28, "...For we are also his offspring...we are the offspring of God." In Luke 3:30, we find Adam referred to as the son of God. Angels are referred to as sons of God. Only creatures who are created directly from the hand of God are said to be sons of God; Jesus Christ being the one exception. He is THE Son of God (Matt. 14:33; 16:16, 27, 54, et al) and is in fact the only begotten Son of God.

The likeness of God refers to the original Christ-likeness. (Eccles. 7:29) We as believers can have this as we walk after the Spirit (Gal. 2:20; Col. 1:27). Man was created to manifest godliness. We can only do that now by being associated with Christ (1Tim. 2:10; 3:16; 4:7&8; 6:3-6; Titus 1:1, etc.)

## [aj] Man was created to reign on earth

In verse 28 we see that Adam and Eve (note – "them") and the entire race were to have dominion over the earth. This speaks of a kingdom. Matthew 25:34 speaks of this kingdom having been "prepared for you from the foundation of the earth." This was given to Adam but he lost the right to reign. It was given again to Noah (Gen. 6:18; 9:9) but he failed. It was next given to Abraham and his seed (Gen. 15:21; 17:4-9; 17:19-21) but Israel failed. It was given by God to Nebuchadnezzar, the king of Babylon (Jer. 27:6) when the times of the Gentiles began. It will return to Israel when the Kingdom of Heaven is finally set up (Num. 24:19). It will be through Christ that the Kingdom will be finally established.

## [ak] Male and Female

"In the image of God created he <u>him</u>, male and female created he <u>them</u>…" Both male and female are created in the image of God. Adam was created first, and then Eve was taken out of Adam (Gen. 2:22). Though that image was marred and effaced by sin, believers today can again find that image "in Christ" (Eph. 4:24; Col. 1:15-20) who is "the image of the invisible God." God will one day conform every believer of every dispensation to the image of Christ (Isa. 43:6-7; Rom. 8:29) once again and will have it so throughout eternity.

God tells us in Malachi 2:14-15 why God created man male and female: "…That he might seek a godly seed. Therefore take heed to your spirit, and let none deal treacherously against the wife of his youth." This is God's purpose statement for marriage.

## [al] First Blessing of the Bible is in verse 28

It was not until God brought animal life on the earth that God "blessed" a creation of his (at least in the written record). God blessed the fish and fowl (Gen. 1:22), the animal life and man (Gen. 1:28), and he blessed the seventh day. In Genesis 5:2 we find that God specifically blessed Adam and Eve (calling their name Adam) in the day they were created. In Genesis 9:1 we find God similarly blessing Noah and his sons.

## [am] Adam's Commission

Here in verse 28 we have a five-fold commission given to Adam and Eve:

1. Be fruitful – Not just have kids, but being productive for your purpose in creation is to raise godly children.
2. Multiply – Increase in number.
3. Replenish the Earth – It was once "plenished" (cf. Gen. 9:1; Isa. 2:6; 23:2; Jer. 31:25) but was depopulated and now had to be repopulated (replenished).
4. Subdue it – It was not in a subdued state but Adam was to win it back. Man was given custodial charge of the earth (Hebrews 2:7)
5. Have Dominion over it – Establish a kingdom; reign over it as regent for God.

## Genesis 1:29-31 The Green Herb given for Life on Earth

²⁹ And God said, Behold, I have given you every herb[an] bearing seed, which *is* upon the face of all the earth, and every tree, in the which *is* the fruit of a tree yielding seed; to you it shall be for meat. ³⁰ And to every beast of the earth, and to every fowl of the air, and to every thing that creepeth upon the earth, wherein *there is* life, *I have given* every green herb for meat: and it was so. ³¹ And God saw every thing that he had made, and, behold, *it was* very good. And the evening and the morning were the sixth day.

## [an] Man's Diet

The original diet for both man and beast was vegetative. This passage speaks of suitable food being:

1. Cereal grains – "herb bearing <u>seed</u>"
2. Fruits and nuts – "every tree in the which is the <u>fruit of a tree</u>"
3. Fleshy and leafy vegetables – "every <u>green herb for meat</u>…"

All of this was changed after the flood (Gen. 9:3-4) when men and beasts (see Psa. 104:21) could now eat the flesh of animals. "Every moving thing that liveth shall be meat for you; even as the green herb have I given you all things. But the flesh thereof with the life thereof, which is the blood thereof, shall ye not eat."

Still later, God limited the diet of the children of Israel under law to "clean" animals (Lev. 11:47). Today, under the dispensation of the grace of God, we may eat anything. "For every creature of God is good, and nothing to be refused, if it be received with thanksgiving…" (1Tim. 4:4).

In the kingdom, the animal kingdom will again revert back to a vegetative diet (Isa. 11:6; 65:25).

## [ao] Man was created as a Tri-part creature

Man was created in the image and likeness of God in order to relate to God and to think like God.

The Image of God:-- Tri-Part makeup of man as body, soul and spirit is evident from passages such as 1Thessalonians 5:23. In our Bodies we each have a Brain but that is not all there is to the thinking part of man. In our Souls we each have a mentality that the Scripture calls the <u>Heart</u> (Romans 10:9 and 10). This is the seat of our emotions and is what constitutes our personality. In our Spirits we each have a mentality that Scripture calls the <u>Mind</u> (1Cor. 2:16). The Spirit: Gives us God consciousness. The spirit enables us to relate to God (God's Spirit witnesses with our spirit… - Romans 8:15-17).

**The Spirit:** Gives us God Consciousness and enables us to receive information from God.
- The spirit enables us to understand the things of God.
- The "Mind" as we find it in scripture is the mental attitudes (the mentality) of the human spirit.
    o Enable us to serve God (Rom. 7:25).
    o Can be carnal (Rom. 8:7).
    o Is renewed by the Scriptures (Rom. 12:2; Eph. 4:23).
    o Can be the mind of Christ (1 Cor. 2:16) as we study the Word of God.
    o Is known by the Holy Spirit (Rom. 8:27).

**The Soul:** Gives us Self consciousness. It is our identity. It is who we are – our real person (Matt. 12:34; Prov. 23:7).

The "Heart" as we find it in scripture is the mental attitude (the mentality) of the soul.

The Heart is where we hold our convictions "with the heart man believeth unto righteousness" Rom. 10:10.

The beliefs and convictions of the heart produce our emotions. The heart can:

Grieve (Gen. 6:6)
Communicate (Gen. 17:17)
Be glad (Exd. 4:14)
Be hardened (Exd. 4:21)
Be sincere (Num 21:39)
Be discouraged (Deut. 1:28)
Fear (Deut. 7:17)
Love (Deut. 6:5).

But the emotions of the heart can not "think" independent of the spirit. Our emotions follow the thinking done based on information that we store in our spirit (in our mind as we saw above).

Therefore: The heart can be deceived (Deut. 11:16). Yet it is the soul that decides what we each will do with our body and with our spirit (1Cor 6:20).

The Soul and Spirit can exist without the body. We see in Luke 16:23 where we see the rich man, Lazarus, and Abraham in their disembodied state.

**The Body:** Iis that part of us that enables a person to carry out the decisions that he or she makes in his or her soul. We can live in a physical environment because we have a body.

Genesis Study Guide Questions
Chapter 1

1. Genesis 1:1 says simply "In the Beginning…" What exactly according to this verse was created in the beginning? Don't go to the rest of the chapter to answer this. Stay in this verse.

2. According to John 1:1-10, who was there in the beginning? Who was it according to John 1:3 that made all things that have been made?

3. According to Hebrews 11:3, it is by (_____) that we understand that the worlds were framed by the Word of God. What is the name given to Jesus in John 1:1?

4. Two different Hebrew words are used in Genesis Chapter 1. One is "bara" translated "created" and the other "basah" which is translated "made." What is the difference between the two? List things that were created in Chapter 1. List things that have been made. Do you see a significant difference between things created and things made?

5. Colossians 1:16-20 talks about the heaven and the earth. Does that passage give us a clue as to the two fold purpose for the reference to heaven and earth in Genesis 1:1?

6. Genesis 1:2 says "And the earth was without form and void…" We noted that the word translated "was" is also translated "became" and "become" in Genesis 2:7; 18:18; and 19:26. We can understand this to mean that the earth was not created without form and void but became that way. We saw that the words "without form and void contain the Hebrew words "Tohu" and Bohu". We noted from Isaiah 45:18 we read that "he created it not in vain (tohu)" but rather that he created it to be inhabited. If the earth were inhabited before man was created to inhabit it, who would have been the ones who inhabited it?

7. The deep covered the earth in Genesis 1:2. It will again cover the earth in Genesis 7:11 and 8:2. In both cases, the deep was used to mete out judgment. According to Job 26: 7-13; 38:30; and Psalm 104:6 the deep is still there. Where is it (at least the upper part of it)?

8. In John 8:44 we noted that Satan is referred to by the Lord as being "a murderer from the beginning." We note that the Greek word translated "murderer" is the word for a "man slayer." But we read in Ezekiel 28:15 that Lucifer (who became Satan) was "…perfect in [his] ways from the day [he] was created until iniquity was found in [him]." Obviously he was not a murderer from his beginning but from man's beginning. Since we do not find an account of the creation of Lucifer from Genesis 1:3 through the end of Chapter 1, what must we conclude about the time of his creation in the context of Genesis Chapter 1?

9. Between what two verses in chapter 1 would we place the creation and fall of Lucifer? Would we then put the events of Isaiah 14:12-15 and Ezekiel 18:11-19 in this gap?

10. If Lucifer (Satan) was not a murderer from his beginning, then he was a murderer (man slayer) from whose beginning?

11. If Lucifer was the anointed cherub that covered the throne, and he was in Eden, then where must the throne of God have been between Genesis 1:1 and 1:2?

12. If step one in Lucifer's rebellion against God is that he would ascend into heaven, where is it that he was not to be? Remember that according to Genesis 1:1 there were only two places where he could have been. How many heavens are spoken of in Genesis 1:1 in your KJV?

13. Where according to 2Corinthians 12:1-2 is the throne of God today. Why would God have moved His throne off from the earth to the third heaven?

14. Why would it be that from Genesis 1:2 on in the Bible we don't find much reference to the heavens until we come to the Pauline Epistles?

15. How many separate acts of God do we see in Genesis Chapters 1 and 2?

16. According to Job 38: 4-11, who was there to watch the operation when God "laid the foundations of the earth?" Who would those "morning stars" and "sons of God" have been?

17. In what verse in Genesis Chapter 1 does the actual seven days of creation week start?

18. Lay out for me the events of what happened on each day of creation week. Start the week on Sunday morning and define the itinerary of each day of the week.

19. What did God make to occupy the open space that was created by the dividing of the waters that covered the earth in Genesis 1:2?

20. According to verse 16, how many lights were actually "made" on the fourth day? What lights beside the moon were to rule the night? Beside the lights that were made on the fourth day of creation week, how many different lights are referenced in verse 16? When were the stars created?

21. In verse 26 we see God making man in His image, after His likeness. In what sense is man in the image of God? In what sense is he in God's likeness?

22. According to Genesis 1:28, what was man to do with God's creation on earth? List the fivefold commission that God gives man in this verse.

23. What according to verses 29 and 30 was to be man's (and animals') diet?

## The Generations of the Heaven and the Earth

### Genesis 2:1-3 (KJV) - The Sabbath of Rest

[1] Thus the heavens and the earth were[a] finished, and all the host of them. [2] And on the seventh[b] day God ended his work which he had made; and he rested on the seventh day from all his work which he had made. [3] And God blessed the seventh day, and sanctified it: because that in it he had rested from all his work which God created and made.

### [a] "The heavens and the earth were finished" Verse 1

Thus the heavens and the earth were finished, and all the host of them [i.e. of the heavens and the earth]. Note that in Genesis 2:1 the heavens are plural while in Genesis 1:1, we saw that "In the beginning" God created the heaven (singular) and the earth. In creation week, God added the first heaven (the earth's atmosphere) and the second heaven – the stellar universe (though this existed as part of the original creation, it was from Genesis 1:7 on that there was a separation of the second from the third heaven by the "waters that are above the heavens" (Psa. 148:3-5)

All the host of the heaven (angelic) and the earth (animal and human) were finished in Genesis Chapter 1. See Exodus 20:11; 31:17; 2 Kings 19:15; 2 Chronicles 2:12; Psalm 89:11-15; 104:12; 136:59. "Thou, even thou, art LORD alone; thou hast made heaven, the heaven of heavens, with all their host, the earth, and all things that are therein, the seas, and all that is therein, and thou preservest them all; and the host of heaven worshippeth thee." (Neh. 9:6)

"But ask now the beasts, and they shall teach thee; and the fowls of the air, and they shall tell thee: or speak to the earth, and it shall teach thee, and the fishes of the sea shall declare unto thee. Who knoweth not in all these that the hand of the LORD hath wrought this? In whose hand is the soul of every living thing with the breath of all mankind." (Job 12:7-10)

The "host" of the heavens includes the sun, moon and stars (Deut. 1:19; 17:3; Psalm 33:6, 9; Jer. 8:2; etc.). Isaiah 34:4 speaks of a time when all the host of heaven shall be dissolved. This is a reference to Revelation 21:5 when God makes all things new.

### [b] The Seventh Day Verses 2 &3

The seventh day is significant in Scripture. Here we see that God "ended his work which he had made. And God blessed the seventh day and sanctified it because that in it he had rested from all of his work which God created and made." This is not rest required by exhaustion but rather the rest of one who has finished work that he is proud of and which he looks forward to enjoying. We will see in Genesis 3:18 that other things are added to God's creation: thorns and thistles. These are the painful, unpleasant negative elements. They were not a part of the creation as completed in Genesis 1. The statement in Genesis 2:2 is a statement of satisfaction; i.e. creation was the way God wanted it. It is interesting to note that from this time forward in scripture, the number seven is the number for perfection and completion. The seventh day resulted in a completed and perfect creation. However, sin entered into that perfect creation so God entered into His work of redemption which will take millennia to complete -- Perhaps seven millennia.

God did three things in connection with the seventh day:
1. He rested! This speaks of His satisfaction with His work. At the end of the sixth day, everything (so far) was the way he wanted it. This is the satisfaction that a man would have who had just finished building the house of his dreams. The rest would be the rest of enjoying what he had made. This

rest, we will see, will be broken in Chapter 3 by an enemy. However, at this point in time, God looked back at what He had made "...and, behold, it was very good" (Gen. 1:31). There is a dispensational significance to the seventh day (i.e. the Sabbath). It is as Hebrews 4:9 says, "There remaineth a rest for the people of God." Though the rest of Genesis 2:2 was broken by sin, there is a coming day when God and man will again rest when everything will again be perfect and no sin will ever again enter to break that rest. "And there shall in no wise enter into it anything that defileth neither whatsoever worketh abomination, or maketh a lie: but they which are written in the Lamb's book of life." (Rev. 21:27)

It should be noted that this is the only Sabbath that the Lord has ever observed. Man's sin has broken the Sabbath of rest for God. Ever since then, God has been at work repairing that broken rest. "My Father worketh hitherto and I work" (John 5:17). A Sabbath can only be truly celebrated when there is nothing to be done. Our Lord while on earth had no rest. The Sabbath of the law was not given to God but to man. He is the Lord of the Sabbath; the one who owns the Sabbath (Matt. 12:8). That is because the Sabbath was made for man (Mark 2:27). The fact that Christ worked on the Sabbath (Luke 6:8) is proof that He is God much as was His power to forgive sins (Mark 2:5-11; Luke 5:20-21). Christ was in the grave working redemption while man was celebrating the Sabbath. The Sabbath was never changed (Matt. 28:1). The Sabbath belonged to the law and to Israel. The church which is Christ's body is free from the law and the Sabbath being as free as Christ Himself (Col. 2:10). We do not keep the day that our Lord spent in the tomb. Rather, Christians traditionally gather on the first day of the week (1 Cor. 16:2; Acts 20:7). While the Sabbath continued to be the meeting day for Israel (Acts 13:14; 44), Paul attended meetings on the Sabbath (Acts 17:2) but told believers to remain free from Sabbath keeping (Col. 2:16).

Note from Matthew 28:1-6 that it was on "end of the Sabbath" that our Lord arose from the grave (cf. Luke 24:1; John 20:1, 19, 26). So we ask "Did God change the Sabbath from the seventh to the first day?" No! There is no statement to this effect anywhere in Scripture. We find a tradition of the first day meeting (1 Cor. 16:2; Acts 20:7) but no command for we are free from the law (Col. 2:14-16). Fellowship is indispensable in a believer's life and it is good to have a day devoted to fellowship. We gather for the purpose of mutual appreciation of faith and Christian practice rather than mere duty and obligation. The Sabbath was given to Israel and Israel will one day have her Sabbath (Heb. 4:9 cf. 3:11). We living today in the dispensation of Grace can have rest for our souls as we assume our position of being seated with Christ far above principalities and powers (Eph 4:10). However, though we look forward to rest at the rapture (2Thess. 1:7), we have no rest in our spirit as we labor to reach others in this present evil world (Gal. 1:4; 2Cor. 2:13; 2:5). The Sabbath was not given as a command until Sinai (Neh. 9:13). We see it given in Exodus 20:11 as a "sign" between God and Israel. It was the sign of the Mosaic covenant.

2. God blessed the seventh day. A blessing from God is a promise that it would be fruitful. God has reserved a Sabbath of rest for Himself and His people (Heb. 4) in which all can rest forever.
3. God Sanctified it. God set that day apart as special. God is looking forward to the time when "the last enemy that shall be destroyed is death."(1Cor. 15:26)

## Genesis 2:4-7 The Generations of the Heaven and the Earth

[4] These *are* the generations[c] of the heavens and of the earth when they were created, in the day that the LORD[d] God made the earth and the heavens, [5] And every plant of the field before it was in the earth, and every herb of the field before it grew: for the LORD God had not caused it to rain[e] upon the earth, and *there was* not a man to till the ground. But there went up a mist from the earth, and watered the whole face of the ground. [7] And the LORD

God formed man *of* the dust of the ground, and breathed[f] into his nostrils the breath of life; and man became a living[g] soul.

## [c] The Generations of the Heavens and the Earth (Vs 4)

The term "These are the generations…" occurs 14 times in the Bible; 11 times in Genesis. This is the first occurrence. This becomes an outline of the Book of Genesis. The term applies to the beginning of a lineage. Here it is a reference to the beginning of the heavens and the earth as we know it.

1. The Generations of the heavens and the earth – Genesis 2:4
2. The Generations of Adam – Genesis 5:1
3. The Generations of Noah – Genesis 6:9
4. The Generations of the Sons of Noah – Genesis 10:1
5. The Generations of the Families of the Sons of Noah - Genesis 10:32
6. The Generations of Shem – Genesis 11:10
7. The Generations of Tereh – Genesis 11:27
8. The Generations of Ishmael – Genesis 25:12
9. The Generations of Isaac – Genesis 25:19
10. The Generations of Esau – Genesis 36:1
11. The Generations of Jacob – Genesis 37:2
12. The Generations of Aaron and Moses – Numbers 3:1
13. The Generations of Pharez (to David) – Ruth 4:18-22
14. The Generations of Jesus Christ – Matthew. 1:1

## [d] God as LORD (Verse4)

The title of God is different in Chapter 2 than in Chapter 1. In Chapter one, we find "God" [Elohim] the creator. In Chapter two, when man appears, we find "the LORD God" [Jehovah Elohim]. This is His title as Redeemer and Savior (a reference to Jesus Christ the Son). Note the usage in the following passages relates to the different roles:

Genesis 7:6 "…as God [Elohim] had commanded him; and the LORD [Jehovah] shut him in…"

1 Samuel 17:46-47 "That all the earth may know that there is a God [Elohim] in Israel. And all this assembly may know that the LORD [Jehovah] saveth."

2 Chronicles 18:31 "Jehoshaphat cried out, and the LORD [Jehovah] helped him; and God [Elohim] moved them to depart from him" Jehovah took care of the saint but Elohim acted on the heart of the uncircumcised Syrians. In the second to last chapter of the Bible, we see Christ as Redeemer. In the last chapter of the Bible, we see Christ as Creator.

## [e] The Environment of Eden (Verses 5 & 6)

God had not caused it to rain on the earth (vs. 5). But there "went up a mist from the earth, and watered the whole face of the ground." (vs. 6) There was not a hydrologic cycle as we know it today. Rather, there was a vapor cycle. Dew was a type of precipitation would condense at night and dissipate during the day. This could be possible if the earth were enveloped in a thick vapor canopy. Such a vapor canopy may have existed from the times of the dividing of the waters until the flood of Noah's day. If so, the vapor canopy would have existed at the top of the earth's atmosphere. The vapor canopy would have accounted for some of the phenomena described in the book of Genesis. Some of the effects might have been:

1. Global Greenhouse
    a. Transmit incoming solar radiation
    b. Retain and disperse some of the reflected radiation from the earth's surface.

  c. Therefore, the earth would have maintained a relatively uniform pleasant temperature over most (if not all) of the earth.
2. Uniform Temperature
  a. Therefore, no rapid movement of air masses.
  b. Therefore, no violent storms.
3. With no global air circulation there would be no hydrologic cycle; no rain.
4. With no dust movement in the upper atmosphere, there would be no means to precipitate the vapor canopy.
5. The relative humidity would have been maintained in a very stable state by daily evaporation and condensation.
6. There would be lush vegetation the world over with no extremes of equatorial heat or polar frigidity.
7. The vapor canopy would filter out the destructive ultraviolet and cosmic radiation. Therefore, there would be very little mutation.
  a. Somatic mutations weaken the individual.
  b. Genetic mutations weaken the species.
8. There would be the healing effect of greater atmospheric pressure.

The vapor canopy, if it existed, would not have constituted the "waters that were above the firmament." The waters that were above the firmament would have had to be beyond the sun and the moon because they were made in the open firmament of heaven (Gen. 1:14-15). In Genesis 7:11, we see that the flood waters came because "…then were all of the fountains of the great deep broken up, and the windows of heaven were opened." So too, when the flood waters stopped, it was because "The fountains also of the deep and the windows of heaven were stopped, and the rain from heaven was restrained. And the waters returned from off the earth continually…" (Gen. 8:2-3)

## [f] The Itinerary of the Sixth Day

Here we see the details of what took place on the sixth day relative to the creation of man:
  "And the LORD God formed man of the dust of the ground" in reference to his Body.
  And "Breathed into his nostrils the breath of life" in reference to his Spirit
  And "Man became a living soul" in reference to his essential being -- his Soul.

## Genesis 2:8-15 The Garden in Eden

[8] And the LORD God planted a garden[g] eastward in Eden; and there he put the man whom he had formed. [9] And out of the ground made the LORD God to grow every tree that is pleasant to the sight, and good for food; the tree[h] of life also in the midst of the garden, and the tree[h] of knowledge of good and evil. [10] And a river went out of Eden to water the garden; and from thence it was parted, and became into four heads. [11] The name of the first *is* Pison: that *is* it which compasseth the whole land of Havilah, where *there is* gold; [12] And the gold of that land *is* good: there *is* bdellium and the onyx stone. [13] And the name of the second river *is* Gihon: the same *is* it that compasseth the whole land of Ethiopia. [14] And the name of the third river *is* Hiddekel: that *is* it which goeth toward the east of Assyria. And the fourth river *is* Euphrates. [15] And the LORD God took the man, and put him into the garden of Eden to dress it and to keep it.

## [g] The Garden of Eden (Verse 7)

"And the LORD God planted a garden eastward in Eden." Note: It is the LORD who planted this garden. It is located in the eastern portion of an area called Eden. The LORD put Adam in Eden. We can get a general location of where Eden was by comparing some Bible passages. Genesis 4:16 – Nod is east of Eden; 2 Kings 19:12; Isaiah 37:17; 51:3. In Amos 1:5 we see that Damascus is in Eden. The area encompasses at least a triangular piece from Egypt and the Nile thence eastward to about where the Tigris River enters the Persian Gulf and thence to northwest to about the head waters of the Tigris and Euphrates rivers and thence southwest along the eastern shore of the Mediterranean Sea. All of the empires and kingdoms (the Assyrian, the Babylonian, the Persian, the Greek, etc.) that we find in the books of the Old Testament were centered in this area.

### [h] Two Trees in Eden (Vs 9)

Two trees are noted of all the trees in the garden:
1. The Tree of Life
2. The Tree of the Knowledge of Good and Evil

The Tree of Life is a tree whose leaves and/or fruit, if eaten, would enable man to life forever. Note Genesis 3:22 "…Lest he put forth his hand, and take of the tree of life, and eat, and live forever…" We see in Ezekiel 47:12 that this tree will be for medicine in the millennial kingdom (cf Rev. 2:7; Deut. 32:8). We see in Revelation 22 verses 2 and 14 that this tree will also be for the healing of the nations in the eternal kingdom. This tree bears 12 manner of fruit which is born on a monthly cycle. It is located today in "the paradise of God" which is today in the third heaven. (2Cor. 12:2).

The Tree of Knowledge of Good and Evil is the tree that bore the forbidden fruit. It was the knowledge of good and evil because they would gain a conscience when they violated God's command by eating of it. Adam and Eve were created innocent. That is to say that they knew neither good nor evil. However, as soon as they would violate a direct command of God, they would be sinners and would immediately have the consciousness of that fact. There is a continuum of wickedness to righteousness on which innocence is in the neutral point. Man started in innocence, but, with the entrance of sin, moves toward wickedness. God provides a redemption that brings man to Himself. God is intrinsically righteous. The redemption that God provides for man brings him, not back to being innocent, but to "the righteousness of God."

### Genesis 2:10-15 Four Rivers in Eden

> [10] And a river[i] went out of Eden to water the garden; and from thence it was parted, and became into four heads. [11] The name of the first is Pison: that is it which compasseth the whole land of Havilah, where there is gold; [12] And the gold of that land is good: there is bdellium and the onyx stone. [13] And the name of the second river is Gihon: the same is it that compasseth the whole land of Ethiopia. [14] And the name of the third river is Hiddekel: that is it which goeth toward the east of Assyria. And the fourth river is Euphrates. [15] And the LORD God took the man, and put him into the garden of Eden to dress it and to keep[j] it.

### [i] The Well Watered Garden (Vss. 10-14)

Here we see a river that flows out of the country of Eden to water the garden. Only here the river parts to form four rivers. This is contrary to the natural formation of rivers where numerous heads come together to form one river. The country of Eden can be roughly located by comparing verses as Genesis 13:10; Amos 1:5; Isaiah 51:3; and 2 Kings. 19:12. It apparently encompasses Egypt (the Nile), Zion (Jerusalem), Damascus, the Tigris (Hiddekel – see Dan. 10:4), and the Euphrates rivers. This would be a triangular area from Egypt, along the eastern shore of the Mediterranean to Ararat, and thence southeast to about Babylon.

**[j] Adam – the Gardener (Vs 15).**

The garden is apparently of interest to God as more than simply a place for man to enjoy. A garden is the meeting place where a king retires to rest (Eccles.2:5).

An interesting comparison can be made of Judges 9:7-14 with the trees mentioned or alluded to in Genesis Chapters 1 thru 3:

> Tree of Life – the Olive
> Tree of Knowledge of Good and Evil – the Vine
> Religion – the Fig tree that provided the inadequate covering for Adam and Eve.
> Apostasy – the Bramble (the Thorns and Thistles that came as a result of Adam's and Eve's sin)

## Genesis 2:16-17 Man was free to eat of any tree but one

> [16] And the LORD God commanded the man, saying, Of every tree of the garden thou mayest freely eat: [17] But of the tree of the knowledge of good and evil, thou shalt[k] not eat of it: for in the day that thou eatest thereof thou shalt surely die.

**[k] Why this one Prohibition on Adam? (Vs.16)**

The Tree of the Knowledge of Good and Evil is given as a statement to Adam that God had reserved to Himself the right to tell Adam what to do. This action by God is to say to Adam that he was under authority. Adam had "innocence" when he was created but he did not have righteousness. There is a continuum from wickedness to righteousness on which innocence is the neutral point. This tree however, is the tree of the knowledge of good and evil. This is another different continuum. This is a reference to human good and human evil. The "good" is to do "good deeds" (i.e. to be good to others). The "good" is as Jonathan spoke of David (1Sam. 19:4), as Psalm 145:9 speaks of the LORD; or as believers works in Matthew 5:16; 1Timothy 2:10; 5:10, 25; 6:18; 2Timothy 3:17; Titus 2:7, 14; 3:8, 14 or Dorcas' "good works" in Acts 9:36. But the good works can also be done for selfish and wicked purposes as in Matthew 7:22. Satan and his ministers can do good works for wicked purposes (2Cor. 11:15). The word "evil" speaks of an attitude or action that will bring harm to another (cf. Gen. 50:20; Deut. 22:22; 1Sam. 2:9). But it can be brought on by the LORD (Josh. 23:15; 1Kings 14:10) who is "righteous in all his works" (Dan. 9:14). Righteousness and goodness are not different words for the same thing. Righteousness is the quality of being right. Righteousness produces goodness but goodness does not produce righteousness. Salvation involves God imputing righteousness to the believer by grace through faith apart from works so that the believer can today produce true "good works" (Eph. 2:8-10).

It is important to note for future reference the sequence of events in verses 16 through 20 of Genesis 2. The commandment in verses 16 and 17 was given to Adam before the LORD God made Eve from Adam's rib.

## Genesis 2:18-20 -- God made a help meet for man

> [18] And the LORD God said, *It is* not good that the man should be alone; I will make him an help meet[l] for him. [19] And out of the ground the LORD God formed every beast of the field, and every fowl of the air; and brought *them* unto Adam to see what he would call them: and whatsoever Adam called every living creature, that *was* the name thereof. [20] And Adam gave names[m] to all cattle[n], and to the fowl of the air, and to every beast of the field; but for Adam there was not found an help meet for him.

**[l]** Here in verse 18 we see the name LORD God. When we see this form of the name used in the Bible the name is translated from Jehovah Elohim. This would be Jesus the eternal Word – the second person of the Trinity. Here He is making the observation that man needed a helper that met his needs or is suitable for him – his complement. So He forms every beast of the field out of the same ground that He formed Adam from.

He then brought each to Adam to see what Adam would name them. But none of these animals could meet his real needs for companionship.

**[m]** The fact that Adam could name the animals indicated that he could actually relate to them. However, none of them could meet his needs for true fellowship. These animals had a personality (your pet dog or cat or whatever it is has personality) but you can't have true fellowship because it lacks a spirit.

**[n]** The cattle here are the domestic animals that would provide assistance to Adam in his life. These were better than the beast of the field but still not adequate for his needs of a special kind of fellowship that one of his own kind would provide.

### Genesis 2:21-25 – Eve taken out of Adam

21 And the LORD God caused a deep sleep to fall upon Adam, and he slept: and he took one of his ribs[o], and closed up the flesh instead thereof; 22 And the rib, which the LORD God had taken from man, made he a woman, and brought her unto the man. 23 And Adam said, This *is* now bone of my bones, and flesh of my flesh: she shall be called Woman, because she was taken out of Man. 24 Therefore shall a man leave his father and his mother, and shall cleave[p] unto his wife: and they shall be one flesh. 25 And they were both naked, the man and his wife, and were not ashamed.

**[o]** There is much significance to the fact that it was a rib that was taken to form the woman. It was not taken from Adam's head or she would rule over him. It was not taken from his feet or he would rule over her. It was taken from his side so that she would be truly a help meet for his emotional and physical needs yet provide him with a special kind of fellowship. God caused Adam to fall into a deep sleep in which he was essentially unconscious.

The LORD took a rib from Adam to make the woman. It must be noted that He did not create Eve as a separate act of creation. This is still a part of the original creation of man – the human race. It was important for redemption purposes that there is only one human race. Therefore, there had to be one seminal head of the race. That way there needs to be only one Redeemer who can redeem everyone in the race. 1Corinthians 11:7-12 elaborates on this special relationship of the man and the woman:

7 For a man indeed ought not to cover *his* head, forasmuch as he is the image and glory of God: but the woman is the glory of the man. 8 For the man is not of the woman; but the woman of the man. 9 Neither was the man created for the woman; but the woman for the man. 10 For this cause ought the woman to have power on *her* head because of the angels. 11 Nevertheless neither is the man without the woman, neither the woman without the man, in the Lord. 12 For as the woman *is* of the man, even so *is* the man also by the woman; but all things of God.

(1Corinthians 11:7-12)

**[p]** Verse 24 presents the institution of Marriage with the words "Therefore shall a man leave his father and his mother, and shall cleave unto his wife: and they shall be one flesh." Marriage and the family are the basic building blocks for society. Marriage is also presented in Ephesians as a type of the special relationship that every believer has with Jesus Christ today during the Dispensation of Grace.

"24 Therefore as the church is subject unto Christ, so *let* the wives *be* to their own husbands in every thing. 25 Husbands, love your wives, even as Christ also loved the church, and gave himself for it; 26 That he might sanctify and cleanse it with the washing of water by the word, 27 That he might present it to himself a glorious church, not having spot, or wrinkle, or any such thing; but that it should be holy and without blemish. 28 So ought men to love their wives as their own bodies. He that loveth his wife loveth himself. [29] For no man ever yet hated his own flesh; but nourisheth and cherisheth it, even as the Lord the church: [30] For we are members of his body, of his flesh, and of his bones. [31] For this cause shall a man leave his father and mother, and shall be joined unto his wife, and they two shall be one flesh. [32] This is a great mystery: but I speak concerning Christ and the church."

(Ephesians 5:24-32)

Study Guide Questions on Chapter 2

1.  Verse 1 talks about the host of the heavens and the earth. List what this host consists of.

2.  Genesis 1:1 makes reference to heaven as being singular while Genesis 2:1 refers to the heavens (plural). What does this imply about what happened between these two verses?

3.  Psalm 148: 3-5 talks about waters that were above the havens. Where is that water relative to the three heavens? Consider Job 37:18 and Job 38:30 in your answer. Consider also Job 26:9 &10.

4.  God rested on the seventh day according to verse 2. Was this a rest of exhaustion or of satisfaction? List three things that God did on the seventh day.

5.  Is there a dispensational significance to the Seventh Day (the Sabbath)? Consider Hebrews 4:9 and Colossians 2:16 in your answer.

6.  What does the term "the generations of…" mean? List the eleven generations listed in Genesis.

7.  Verses 4 and 5 refer to God as "the LORD God" – as the one who made the earth and the heavens. Compare this to John 1:1-4. Who (which person of the Godhead) is this title referring to?

8.  What was the environment of the earth like at that time according to verses 5 and 6?

9.  If there were a vapor canopy over the earth that filtered the Ultraviolet radiation from the sun, would that possibly be the water that was "above the open firmament that Genesis 1: 14 and 15 speaks of?

10. Verse 7 through 24 lay out the itinerary of the sixth day of creation week in more detail than we saw it is Chapter 1. Write out the sequence of events of the sixth day.

11. In verse 16, God tells Adam of the forbidden fruit. In verse 17 – 20 God creates animals for Adam to name. In verse 21 – 24, God makes Eve from a rib taken from Adam. Is there significance to this sequence?

12. Where on earth was Eden?

13. According to verse 9, there were two trees in Eden that were specifically named. What were they?

14. Why do you suppose that God imposed that single restriction on Adam and Eve?

## Man Meets the Man Slayer

### Genesis 3:1-5 The Serpent (the man slayer) enters the scene

[1] Now the serpent[a] was more subtil than any beast of the field which the LORD God had made. And he said unto the woman[b], Yea, hath God said, Ye shall not eat of every[c] tree of the garden? [2] And the woman said unto the serpent, We may eat of the fruit of the trees of the garden: [3] But of the fruit of the tree which *is* in the midst of the garden, God hath said, Ye shall not eat of it, neither shall ye touch it, lest ye die. [4] And the serpent said unto the woman, Ye shall not[d] surely die: [5] For God doth know that in the day ye eat thereof, then your eyes shall be opened, and ye shall be as gods[e], knowing good and evil.

### [a] The Serpent (Vs 1)

Who is the serpent? He is not a beast of the field for he is more subtle than any beast of the field. Revelation 12:9 and 20:2 clearly identifies the serpent as the devil. The use of the term "serpent" is a figure of speech to indicate his character and mode of attack. He sneaks in quietly and unnoticed to trick and beguile with the ultimate purpose being to destroy.

### [b] The Temptation of Eve

The serpent approached the woman. 1 Timothy 2:14 gives us a clue as to why he did—there is apparently an element of gullibility in the female of the species that is not as prevalent in the male. Also, Eve heard the information regarding the tree of the knowledge of good and evil second hand, indirectly through Adam. Eve was created after that command was given to Adam (cf. Gen. 2:16-25). The problem therefore might have been that Adam did not accurately communicate the information to Eve. Whatever the case was, the ultimate problem was that the truth of the Word of God was not in Eve's mind at the time of the temptation. In Romans 5:14 we find that Adam and not Eve is credited with the transgression regarding the eating of the forbidden fruit.

### [c] Satan Attacks the Word

In verses 1-5 we find illustrated five common attacks on the Word of God:
1. People Question the Word as Satan did in his approach to Eve. He approaches people today in much the same way:
    a. "Did God really say that?"
    b. "Is the Bible really the Word of God?"
    c. "Is it really without error?"
    d. "Did God really preserve it?"
2. People Subtract from the Word of God as Eve did when she left out the words "freely eat":
    a. They leave part of it out.
    b. They take only that part that they want and ignore the rest.
3. People Add to it as Eve did when she added "Neither shalt thou touch it"
    a. They add words into the text that God did not put there.
    b. They add extra-Biblical material and claim it to be from God. Some examples today:.
        i. "Divine Tradition" of Roman Catholicism
        ii. The Book of Mormon of the Latter Day Saints
        iii. The Watch Tower Organization of the Jehovah's Witnesses

    iv.  Mary Baker Eddy with Christian Science
    v.  Ellen B. White of the Seventh Day Adventists
    vi.  Pentecostal "Tongues, Prophecy, and Revelation"

**Important Note:** All of these groups have what they call modern day prophets whereby they claim to have divine revelation from God that is not in the Bible. This is an attack on the Word of God in that they are going to some other authority for what they call "truth" from God. In anticipation of that, the Lord put in the Bible information that guards against such a claim. In Colossians 1:23-26 we read "If ye continue in the faith grounded and settled, and *be* not moved away from the hope of the gospel, which ye have heard, *and* which was preached to every creature which is under heaven; whereof I Paul am made a minister; Who now rejoice in my sufferings for you, and fill up that which is behind of the afflictions of Christ in my flesh for his body's sake, which is the church: Whereof I am made a minister, according to the dispensation of God which is given to me for you, to fulfil the word of God; *Even* the mystery which hath been hid from ages and from generations, but now is made manifest to his saints…" We note from this passage that the apostle Paul had a ministry to complete (to fully fill) the Word of God by adding the body of doctrine called the Mystery to it. That means that the Word of God was then (with the addition of the mystery to it) complete and nothing more was to be added to it. Therefore, anyone who claimed to be a prophet with information from God after that was a false prophet and the material that he (or she) put forth was not to be regarded as a part of the cannon of Scripture.

4. Beside adding to the Word, they Water it down as Eve did when she said "Lest ye die" implying that you might or you might not die: People do that today by:
    a. Instead of letting Scripture speak, they make it an allegory.
    b. Instead of receiving it as literal truth, they make figurative spiritual applications of it. Much of this comes from a failure to rightly divide the Word of truth.
5. People flat out deny the truth of it as Satan finally did with the words "Thou shalt not surly die." This is the ploy of humanism today.
    a. The theory of evolution whereby the world we live in is not the product of a purposeful Creator but is merely the result of time and chance is man's rejection of God's revelation.
    b. Secular humanism is a religion founded on evolution.
    c. Humanism claims that man descended from animals so that man can be excused for living like animals.

## [d] Satan's Deception

Consider Satan's attack and how Eve played into his scheme:
1. He questions God's goodness, "Yea hath God said, ye shall not eat of every tree of the garden?" He gets her to start thinking that God is depriving her of some good thing.
2. Eve subtracts from the Word of God by leaving out "freely." Thus she overlooked God's abundant supply and provision.
3. She adds to what God said by adding "neither shall ye touch it." Thus making God appear spuriously overbearing in her eyes.
4. She waters it down saying "lest ye die" instead of "thou shalt surely die." Thus she adds uncertainty to the Word and ambiguity to the consequences of disobeying it.
5. The Devil now realizes that he has her in that she did not know the Word. He now flat out denies the Word saying, "Thou shalt not surly die."

Side notes:
Having the knowledge of good and evil condemns (Deut. 1:39). Adam was apparently with Eve when she ate of the fruit (Gen. 3:6). It is interesting to note in Numbers 30:6-16 that under the Law (which was not in effect yet), if a woman's husband was to disapprove of the oath that a woman made, then the oath was

disavowed. We wonder what would have happened here if Adam would have approached God with this problem of Eve having eaten of the forbidden fruit instead of going along with her.

### [e] The gods of Verse 5

An important point is noted in verse 6. Adam was apparently (as noted above) with Eve as the conversation went on between Eve and the Devil. The "gods" in verse 5 are the angelic creatures that God had created "In the beginning." It is apparent that Eve could see these angels and relate to them. She also knew that she and Adam were created lower than them (Heb. 2:7; Psalm. 8:5). Adam and Eve could die (they were mortal) but the angels were not mortal. Satan was playing on her self esteem. She saw an opportunity to "better herself" by being like the angels. The main problem here is with Adam; he kept his mouth shut and did not intercede to stop the transaction from proceeding any further. Even after Eve had eaten of the tree, Adam could have gone to God with the problem and had the problem solved (cf. Num. 30:6-16). But here again, we find Adam lacking in the knowledge of the goodness of God. Adam instead imagines that he is facing a decision of either joining Eve in death or living alone with God without her. The rest is history. Adam joined Eve in spiritual death. As with Adam, so it is with men today; it is man's ignorance that gets him in trouble and fixes his destiny (Eph. 4:18).

- Man is ignorant in thought, word, and deed when he is ignorant of God.
- Man is pure in thought, word, and deed when he knows God.
- The knowledge of God in man has, in every dispensation:
  - Given life to the soul
  - Purified the heart of wickedness
  - Given peace and tranquility to the mind
  - Elevated the affections of the soul to godly standards
  - Sanctifies the character and conduct

### Genesis 3:6-8 -- The Threefold Temptation

[6] And when the woman saw[f] that the tree *was* good for food, and that it *was* pleasant to the eyes[g], and a tree to be desired to make *one* wise, she took of the fruit thereof, and did eat, and gave also unto her husband with her; and he did eat. [7] And the eyes of them both were opened, and they knew that they *were* naked[h]; and they sewed fig[i] leaves together, and made themselves aprons. [8] And they heard the voice of the LORD God walking in the garden in the cool of the day: and Adam and his wife hid[j] themselves from the presence of the LORD God amongst the trees of the garden.

### [f] The anatomy of sin and temptation:

The Bible Doctrine of Peccability

Peccability answers the question "when does temptation become sin?"
  When does sin start?
  Five Steps:
  1. The Presentation: An opportunity is set before you by someone who wants to draw you into sin.
  2. Illumination: The thing is pointed out to you as attractive and/or pleasurable.
  3. Debate: You entertain the thought of doing it and enjoying the pleasure that it would bring. Here is where sin starts.
  4. Decision: You decide to do it. It becomes sin at this point.

5. Action: You do it. But sin brings forth death when it is finished (James 1:15). There are "the pleasures of sin" that can be enjoyed "for a season" (Heb. 11:25), but sin can be avoided (2 Cor. 10:5) for the believer by obeying from the heart the doctrine of grace (Rom. 6:17) "But God be thanked, that ye were the servants of sin, but ye have obeyed from the heart that form of doctrine which was delivered you."

## [g] The Threefold Temptation

The temptation of Eve was three-fold and corresponded with the three-part makeup of man. It also corresponds to the three fold temptation of Christ and the victory that He had over sin.

Table 5: The Threefold Temptation of Man

| Three Parts of Man | Three types of Temptation (1John 2:15-16) | Adam and Eve's Temptation and Failure | Christ's Temptation and Victory (Matt. 4:3-8; cf. Heb. 4:15) |
|---|---|---|---|
| Body | Lust of the Flesh | It was good for food | Stones to bread |
| Soul | Lust of the Eyes | It was pleasant to the eyes | All the kingdoms of the world |
| Spirit | Pride of Life | A tree to make one wise | Cast thyself down |

Christ was tempted in all ways that we are yet He never sinned. Where Eve failed by not knowing the scripture, our Lord succeeded by knowing it, believing scripture, and quoting it from memory. Note how the Lord used scripture to defeat Satan in the threefold temptation of Christ by the Devil in Matthew 4:1-10.

## [h] They Knew They Were Naked (Vs 7)

Adam and Eve both knew that they were naked the instant that they sinned. Apparently they did not know that they were naked before they sinned. It could be that, while they were innocent, they were clothed with an aura of light. They were created in the image of God and God "coverest [Himself] with light as with a garment." (Psa. 104:1). The realization that they were naked is here associated with the eyes of both of them being opened (verse 7). Innocence is the lack of the knowledge of good and evil (Deut. 1:39). Their eyes were open not to the fact that they were naked, but to the fact that they now knew good and evil. This loss of innocence rendered them naked. Adam and Eve now had a conscience—to have the knowledge of good and evil is to have a conscience. Adam and Eve now had the knowledge of good and evil but they did not have the full knowledge of God. True they knew God as Elohim (God) but they did not know Him as Jehovah Elohim (LORD God). They knew God as creator but not as the loving redeemer. Here is where they begin to know Him as LORD God (the redeemer). Satan had succeeded in shaking Eve's confidence in God's love and concern for her and His interest in her welfare. As a result, she traded God's truth for Satan's lie and thus Satan displaced God in her life.

## [i] The covering of fig leaves

Adam and Eve's knowledge that they were naked, as stated above, is their knowledge of good and evil. Innocence is the ignorance of both good and evil.

- Evil is that which is hurtful, harmful, or unpleasant.
- Good is that which is pleasant and benevolent.
- Evil is not synonymous with wickedness (see note [j] of Chapter 2)
- Good is not synonymous with righteousness.

In Adam and Eve, as in us today, the presence of evil brings the consciousness of good. Man, left to himself, will seek to cancel the evil with the good. This is the basis of all man made religions and religious systems. Here, when man finds himself naked, he seeks to cover that nakedness with good—religion as represented by the apron of fig leaves.

## [j] Man hides from God (vs. 8)

The knowledge of good and evil is conscience. Conscience can only make us cowards. Conscience can not bring us to God. Conscience compels us to hide from God. The knowledge of what I am cannot bring me to God unless accompanied by the knowledge of what and who God is.

Conscience produces:
- Shame
- Self reproach
- Anguish
- And most notably—religion

But religion serves only to hide God from our view. Religion is based upon a sense of nakedness. A person must know that he is clothed before he can do anything acceptable to God.

The true believer takes the clothing that God provides. The religionist clothes himself with his own works. The believer works because he is clothed. The religionist works to be clothed.

Had Adam known God's perfect love, he would not have been afraid. "Herein is our love made perfect, that we may have boldness in the day of judgment: because as he is, so are we in this world. There is no fear in love; but perfect love casteth out fear: because fear hath torment. He that feareth is not made perfect in love. We love him, because he first loved us." (1 John 4:17-18).

Adam believed Satan's lie that God did not love him. Adam knew he had sinned and he knew that God and sin cannot go together (Heb. 1:13). As long as there is sin on the conscience, there is distance from God (Heb. 9:14; 10:26). True faith in the redeeming work of God clears the mind of an evil conscience. Creation reveals God's power and Godhead (Rom. 1:20). The cross reveals God's love, holiness, and justice. Grace reveals God's wisdom.

### Genesis 3:9-13 Man hides from the God who seeks him

> [9] And the LORD God called unto Adam, and said unto him, Where[k] *art* thou? [10] And he said, I heard thy voice in the garden, and I was afraid, because I *was* naked; and I hid myself. [11] And he said, Who told thee that thou *wast* naked? Hast thou eaten of the tree, whereof I commanded thee that thou shouldest not eat? [12] And the man said, The woman whom thou gavest[l] *to be* with me, she gave me of the tree, and I did eat. [13] And the LORD God said unto

the woman, What *is* this *that* thou hast done? And the woman said, The serpent beguiled me, and I did eat.

## [k] God seeks the lost man

God's question, "Where art thou?" Proves two things:
1. Man was lost.
2. God seeks man to save him and to have fellowship with him.

## [l] The Loss -- The depth of Adam's fall is striking when we consider what he lost:
- He lost his dominion over the earth (Gen. 1:28).
- He lost his dignity (Gen. 3:10).
- He lost his happiness (Gen. 2:23 cf. 3:12).
- He lost his innocence.
- He lost his purity.
- He lost his peace.
- He lost his very life with God.

And to top it all, he blamed it all on God.

## Genesis 3:14-15 The Serpent cursed by God

[14] And the LORD God[m] said unto the serpent, Because thou hast done this, thou *art* cursed[n] above all cattle, and above every beast of the field; upon thy belly shalt thou go, and dust shalt thou eat all the days of thy life: [15] And I will put enmity between thee and the woman, and between thy[o] seed and her seed; it shall bruise thy head, and thou shalt bruise his heel.

## [m] The First Gospel (verse 15)

The promised seed of the woman (Isa. 7:14) would do for Adam's race what man could not do for himself.
1. Sin had entered the race and it had to be removed. But man could not do it. It was by man that it came.
2. The serpent had taken the kingdom that was to belong to man. It had to be won back but man could not do it. Man had become the serpent's slave.
3. God's claim for the earth had to be met and man was created to make that claim. However, man failed to do that. He had lost God's claim on the earth. Satan had now become the god of this world (2Cor. 4:4).
4. Death had entered the human race and had to be removed. But man could not do that either. Man was spiritually dead.

Adam had to stand aside and could do nothing but listen to the words of hope regarding the promised seed of the woman.

## [n] The curse on the Serpent

This statement: "…upon thy belly shalt thou go, and dust shalt thou eat all the days of thy life" spoken to the serpent leads some to believe that this is a reference to a snake. But it is actually a reference to the devil's continual defeat in the coming conflict between him and the woman (or the woman's godly seed line), between "his seed and her seed." In the conflict of the ages, Satan was going to be "left in the dust" by the seed of the woman. It may appear that he is winning the conflict by the fact that there are more that do not believe God than those that do. However, in the end, it will be those who find their salvation in God's provision for sin who will find the victory in Christ. As believers look around and see ourselves in the minority, we take heart in the faith that in the end it will be God who is triumphant and we will triumph with Him as we accept His salvation by faith and accept His word as truth.

## [o] The seed of the serpent

The serpent has a seed line that ultimately culminates in the antichrist. The seed line of the woman will ultimately culminate in Christ and all who come to God by faith in Him. In the contest between Christ and Satan, the seed of the woman has his heel bruised while Satan has his head crushed. The bruising of the heel took place at the first advent of the Savior and speaks of His death for the sins of the world. The crushing of the serpent's head will take place at the Second Advent.

The seed line:

> Genesis 4:1 Cain ("was of that wicked one") The name Cain means literally "gotten"
> Genesis 4:25 Seth The start of the godly seed line
> Genesis 9:8 Noah
> Genesis 12:1-7; 13:15; 15:4, 13, 18; 17:7-12 Abraham
> Genesis 17:19; 27:17; 26:2-4 Isaac
> Genesis 28:3, 4, 13, 14 Jacob
> Deuteronomy 1:8 The Twelve Tribes
>  (Judah – Gen. 49:10)
> Romans 1:3 David/Christ
> Revelation 12:16 The woman, the remnant of her seed

## Genesis 3:16-19 The curse placed on man

> [16] Unto the woman[p] he said, I will greatly multiply thy sorrow and thy conception; in sorrow thou shalt bring forth children; and thy desire *shall be* to thy husband, and he shall rule over thee. [17] And unto[r] Adam he said, Because thou hast hearkened unto the voice of thy wife, and hast eaten of the tree, of which I commanded thee, saying, Thou shalt not eat of it: cursed *is* the ground for thy sake; in sorrow shalt thou eat *of* it all the days of thy life; [18] Thorns also and thistles shall it bring forth to thee; and thou shalt eat the herb of the field; [19] In the sweat of thy face shalt thou eat bread, till thou return unto the ground; for out of it wast thou taken: for dust thou *art*, and unto dust shalt thou return

## [p] The Curse on the Woman (Verse 16)

God's words to the woman are very instructive; especially in light of modern thought of fallen man (i.e. the woman's liberation movement). Let's consider and understand the terms:

- "The multiplied sorrow…"
  - Death has now entered the human race. She would see the death of some of the children that she would conceive and bare.
- The multiplied conception
  - Because death would take lives from the earth, her fertility would be increased.
- "In sorrow thou shalt bring forth children."
  - This speaks of the pain of childbirth.
  - The pain of delivery would be a reminder to her of the innocence that was lost and the paradise that they had to leave.
- "Thy desire shall be to thy husband…"
  - An enemy has come on the scene (Actually a threefold enemy—the world, the flesh, and the devil). As long as there are enemies, there has to be headship to provide protection from the enemy (1Cor. 15:25; Eph. 5:23). God put a natural desire within the woman to

desire the love and protection of her husband. In the unsaved, this desire is denied under the influence of the world. This denial is a part of what Romans 1:31 speaks of us being "without natural affections…" Paul tells us that this will be particularly true in "…the last days" when "…perilous times shall come…" (2 Tim. 3:1-3).

- "…and he [thy husband] shall rule over thee."
  - ○ Some independence is lost for her. Her husband will now rule over her and be "the head of the woman" (1 Cor. 11:3).

## [q] The Curse on the Man (verse 17-19)

Upon Adam, God meted out a six-fold curse:

1. "Cursed is the ground for thy sake". This is possibly a reference to the institution of the Second Law of Thermodynamics. Had Adam not sinned, he could have lived forever. Therefore, the environment in which he lived would have had to abide forever. The Second Law of Thermodynamics, the decay principle, therefore it could not have been in effect. (See Eccles. 1:2, 3, 13, 14; Rom. 8:20-22). The Second Law of Thermodynamics (the decay principle) will not be in the eternal kingdom in the new heaven and the new earth. If it were, that eternal kingdom would not be able to last forever. We studied this law in the preface to this study.
2. In sorrow shalt thou eat of it all the days of thy life (Job 5:6, 7; 14:1; Psalm 90:10). The earth will not be as plentiful or as productive as it could be and should be.
3. Thorns also and thistles shall it bring forth to thee. Unpleasant things are now going to be added to creation. This would include parasites of all kinds, mosquitoes, thistles, pathogens, etc.
4. He would now eat the herbs of the field; no longer would he be in a garden lush with fruit.
5. He would have to work hard in order to make a living. However, this is also a promise that he would make a living if he worked hard.
6. The earth that he was originally to reign over would now reclaim his body in death.

## Genesis 3:20-24 Man expelled from Eden

[20] And Adam called his wife's name[r] Eve; because she was the mother of all living. [21] Unto Adam also and to his wife did the LORD God make coats[s] of skins, and clothed them. [22] And the LORD God said, Behold, the man is become as one of us, to know good and evil: and now, lest he put forth his hand, and take[t] also of the tree of life, and eat, and live for ever: [23] Therefore the LORD God sent him forth from the garden of Eden, to till[u] the ground from whence he was taken. [24] So he drove out the man; and he placed at the east of the garden of Eden Cherubims[v], and a flaming sword which turned every way, to keep the way of the tree of life.

## [r] Adam's Statement of Faith (Verse 20)

Here is Adam's statement of faith. He called his wife's name Eve, meaning "Living." This was in a faith response to the promise of Genesis 3:15 -- The promised seed of the woman. Adam understood that the promise was of a redeemer who would provide to man the eternal life that had been lost by sin.

## [s] The First Blood Sacrifice (Vs. 21)

The first bloodshed in creation was the blood of innocent animals as a covering for Adam and Eve's sin. It is apparent from this and Chapter 4 that God had instituted a blood sacrifice. It is apparent also that Cain and Abel knew what they were to bring as an offering, when to bring it, and where to bring it. Adam learned what the price of redemption was as he watched God slay two of the animals that Adam had recently named (Gen. 2:20) to cover his sin and his lack of innocence.

We learn from Romans 3:20-25 that the sins of believers of the past (before Calvary) were remitted only because God the Father knew that the real sacrifice that would enable Him to be just in remitting sins that are past (sins that were committed and then remitted during Old Testament time) was His (the Father's) faith in the coming shed blood of His only begotten Son (Rom. 3:25). Adam understood (or at least he learned) that "…without shedding of blood is no remission." (Heb. 9:22) but he did not know that the real sacrifice was that of the coming redeemer. This is the covering for sin that God provides and not man. This covering for sin enabled man to stand before God as a redeemed soul.

## [t] Man Barred from the Tree of Life (Vs. 24)

Adam could have eaten of the tree of life and live forever. He could have freely eaten of this tree before the fall (Gen. 2:16) – there being only one tree that he was forbidden to eat of (Gen. 2:12). Now, if he were to eat of this tree, he would live forever in a fallen state. Therefore, God drove man from the garden and placed cherubim on guard of the east gate of the garden to guard the way to the tree of life. Man cannot again have access to the tree of life until the kingdom is established (Rev. 2:7; 22:2, 14) but that will be after the sin nature is eradicated from man's makeup.

## [u] The New Role for Man

Now, being a fallen creature, with sin and death having entered God's creation, instead of reigning over the earth, Adam labors and struggles with the earth to make a living until the earth wins the struggle and reclaims his body in death.

## [v] The Cherub

The presence of the cherub is significant. The next place in scripture that we see them is at the mercy seat in Exodus 25:18-22. There and every place thereafter, they are associated with the throne of God or the presence of God. It would appear that this then was where the blood sacrifices were to be offered -- at the entrance to the garden where the cherubim stood (Cf. 1Sam. 4:4; 1Kings 6:25-35; Psalm. 80:1; 99:1; Ezek. 10:2; Num. 22:23).

## [w] The Seed of the Woman (Vs. 15)

"The seed of the woman" is a very important term in Scripture. The term "seed" everywhere else refers to a man's seed. But here we see the reference being to the seed of the woman. It must be remembered here that Eve was taken <u>out</u> of Adam <u>before</u> the fall of Adam. There is a coming offspring of the woman who would not and did not have a human father. That one would be born without a sin nature. The sin nature is thus understood to be passed on through the father – through the male line.

## [x] A New Dispensation Begins

Genesis 3:15 begins the Dispensation of Promise. The Dispensation of Innocence ended with Genesis 3:6-7. (See chart below)

## Table 6:
## Dispensations and Institutions of God

| Time Past | | | | But Now | Ages to Come | |
|---|---|---|---|---|---|---|
| **Before the World Began** — God made a promise to Himself of eternal life (Titus 1:2) | **Adam in Eden** — Innocence | **Adam to Moses** — Promise (Mat 25:34) | **Moses** — Law | **Paul** — Grace | **Christ** — Kingdom | **Dispensation of the Fullness of Times** (Eph 1:10) |
| | | | Circumcision Uncircumcision (Ephesians 2:12) | One new man Jew and Gentile saved in one body | Christ reigns on earth over Israel and Israel a blessing to the nations | |
| | Man walked and talked with God | Adam / Adam (sin entered) / Noah / Shem / Abraham / Isaac / Jacob / 12 Tribes / Moses | | | | |
| The Angelic Creation | Adam was to Reign | Death Reigned (Rom 5:14) | Sin Reigned (Rom 5:12, 21) | Grace Reigns Rom 5:20, 21 | Righteousness Reigns | All things reconciled to God (Col 1:20) |

Romans 5:12: By one man sin entered the world and death by sin

> Death passed upon all men for all have sinned (in Adam).
>
> Sin was in the world
>
> But sin was not imputed when there is no law
>
> Nevertheless, death reigned from <u>Adam to Moses</u> even though they did not sin against a direct commandment of God as did Adam -- yet they died.

Galatians 3:19 The law was added to the Promise

> The promise was before the law.
>
> The law was added to the promise.
>
> > The Law did not replace the promise.
>
> The law is interrupted by grace. Romans Chapters 1-3 tells us that Paul was separated unto the gospel… which he had promised afore in the Holy Scriptures.
>
> Law will continue on with the New Covenant
>
> > But then it will be written in their hearts (Jeremiah 31:31)
>
> The Promise is fulfilled in the New Covenant
>
> (But we living in the dispensation of grace have the spiritual benefits of the New Covenant under Grace)

> By Adam came the Entrance of Sin
>
> By Moses came the Entrance of the Knowledge of Sin by the Law
>
> By Paul came the Entrance of the Forgiveness of Sin by the grace of God.

The Institutions of God include: 1) Volition 2) Marriage 3) Family 4) Human Government and 5) the Blood Atonement.

Human Government is an institution, not a dispensation. There are certain operating principles that God instituted by which man could live out the plan and purpose that God had for man. The institution of free will was the first. By this man was given the freedom to choose his own course of conduct but God would hold him accountable for the choices that he would make. The next in order of occurrence was marriage and then family. We will see later God also instituted government after the flood of Noah. We also add to the list of institutions of God the institution of the blood atonement in that God apparently had informed man that the blood sacrifice was required for the covering of man's sin. We will see this later in Chapter 4. Conscience is really part of volition which is an institution and not a dispensation. From Adam onward, Promise was the issue (Matt. 25:34).

## [y] The Dispensation of Promise

Genesis 3:15 begins the Dispensation of Promise. Following the seed line will eventually lead us to Christ who is the real promised "…seed of the woman…" Christ is the one who will crush the serpent's head. The history of the Old Testament from this point on is then a continuing conflict between God and Satan wherein Satan attempts to destroy the seed line so as to prevent the fulfillment of the promise of Genesis 3:15.

## [z] The Headship of the Man

The headship of the husband is seen in Genesis 3:16 as an ordinance of God. Prior to that, they were co-regents under Adam's leadership. Eve usurped Adam's position of leadership and plunged the world (of man) into sin. The real problem was Adam's (because he hearkened unto the voice of his wife). Before the fall, he had failed to be her protector. After her fall, he could have functioned as a priest and interceded for her (Num. 30:5-8). There is a natural tendency for women to usurp authority over her husband and a natural tendency for a husband to let her do it. But, to be the spiritual leader that God would have us be, husbands need to lovingly overcome both of these tendencies.

It is important to note, however that, when people get saved (being justified by faith), God reverses the effect of the fall. In a godly home where both the husband and wife are filled with the spirit of God, the wife willingly submits to the authority of her husband with:

a. A thankful spirit.
b. A joyful spirit.
c. A submissive spirit.

Her husband in turn loves her, honors her, protects her, and provides for her. The man who will not rule his home for the good of all in it will ruin it.

## [aa] The Curse Born by the Savior

It is interesting to note as we study the curse upon the man that our Lord bore the curse himself.

1. Adam had to work around the thorns and thistles. Our Lord had to wear a crown of thorns (Matt. 27:29).
2. Adam would make a living by the sweat of his brow. Our Lord sweat drops of blood as he prayed in Gethsemane (Luke 22:4).
3. Adam would face physical death as a result of his sin. Our Lord faced physical death to rid the world of sin (John 10:18).
4. The ground was cursed for Adam's sake. Our Lord bore the curse (Gal. 3:13).
5. It would be in sorrow that Adam would till the ground. The Lord was a man of sorrows and acquainted with grief (Isa. 53:3).

## [ab] Tracking the Promised Savior

There are seven promises of the Savior in the Old Testament

1. He will be of the human race. (Gen 3:15)
2. He will be descended from Shem (Gen. 9:6)
3. He will be of the Hebrew nation (Gen. 12:3)
4. He will be of the tribe of Judah (Gen. 49:10)
5. He will be of the Family of David (2 Sam. 7:16)
6. He will be born of a virgin (Isa. 7:14)
7. He will be from Bethlehem (Mic. 5:2)

## End Notes to Chapter 3 There is an important tie between Genesis 3:15 and Roman 3:24-26

The promised "Seed of the Woman" in Genesis 3:15 is Our Lord Jesus Christ. Exactly how the seed of the woman would crush the serpents head had to have been kept a secret (a mystery) until it was accomplished. The apostle Paul tells us why in 1 Corinthians 2:7-8 "[7] But we speak the wisdom of God in a mystery, *even* the hidden *wisdom*, which God ordained before the world unto our glory: [8] Which none of the princes of this world knew: for had they known *it*, they would not have crucified the Lord of glory." The real meaning of the promise made in Genesis 3:15 was not revealed in scripture from Genesis through the entire Old Testament. In fact, it is not revealed in the gospel accounts of Matthew, Mark, Luke and John or even in the Book of Acts. It is not until we come to Chapter 3 of the Book of Romans that we find full impact of Calvary. It was on the cross and by the cross that the "promised seed of the woman" actually crushed the serpent's head. Let's study that passage to pick up in the full impact of Calvary.

Romans 3:23-26

"²³ For all have sinned, and come short of the glory of God; ²⁴ Being justified freely by his grace through the redemption that is in Christ Jesus: ²⁵ Whom God hath set forth *to be* a propitiation through faith in his blood, to declare his righteousness for the remission of sins that are past, through the forbearance of God; ²⁶ To declare, *I say*, at this time his righteousness: that he might be just, and the justifier of him which believeth in Jesus."

Every member of Adam's fallen human race except for the one who is the creator has sinned according to verse 23. But Christ Jesus who "knew no sin" (2Cor. 5:21) was made to be sin for us (for us who did the sinning) so that we might be made the righteousness of God in Him. There was an amazing spiritual transaction that took place on Calvary. The sinless Creator was made to be sin by the Father so that He could pay the debt that human sin incurred. At the same time, the righteousness of the Savior is credited to the believer's spiritual bank account.

In verse 24 we read: "Being justified freely by his grace through the redemption that is in Christ Jesus..." The believer today is "justified freely." The word translated "freely" here has the idea of "without a cause" and is translated so in John 15:25 where we read of Christ "...they hated me without a cause." There was no just cause for anyone to hate our Lord but they did hate him unjustly. God had no cause to justify believers but yet He does so by grace. The term "Being justified by grace..." tells us that it is only by the unmerited favor of God that anyone is justified today. The word grace has the idea of delight. God delights in saving sinners. Justification today is through the redemption that is in Christ Jesus. The word for redemption here carries the idea of setting one free from bondage by the payment of a price.

"Whom God hath set forth *to be* a propitiation through faith in his blood, to declare his righteousness for the remission of sins that are past, through the forbearance of God..." in verse 25 is looking back through Old Testament times. It is to show how it was that God the Father could remit the sins of the believers when they brought the blood sacrifice on the basis of the Father's faith in the coming shed blood of the Son. The "sins that are past" refers to the sins of the Old Testament saints. God set Christ forth to be propitiation through faith in is blood. The word translated "propitiation" is translated "mercy seat" in Hebrews 9:5. It has the idea of a fully satisfying payment. The blood applied to the mercy seat in the Old Testament enabled God to remit sins of the Old Testament saints because the Father had faith in the coming shed blood of His Son. It is the blood of Jesus Christ that really gets the job done.

In verse 26 "To declare, *I say*, at this time his righteousness: that he might be just, and the justifier of him which believeth in Jesus..." refers to us who live in the dispensation of grace since the truth of what was accomplished on Calvary was been revealed. God the Father is righteous and just in justifying the believer today because He did not just overlook sin. He cleared the account of sin's debt by the redeeming work of Jesus Christ the Son.

God declared two things when He set Christ forth as the fully satisfying payment for sin. First He declared His righteousness for the remission of sins that are past through the forbearance of God. Secondly He declared His righteousness that He would be both just and He would also be the justifier of sinners. The sins that "are past" here are those committed in Old Testament time before the truth of what the cross accomplished was revealed. This includes sins committed under the Old Covenant "...that by means of death, for the redemption of the transgressions that were under the first testament, they which are called might receive the promise of eternal inheritance" (Hebrews 9:15). "...It is not possible that the blood of bulls and of goats should take away sins" (Hebrews 10:4). Therefore, God's justice demanded that, the blood of the redeemer be shed to truly take away the sin. God could not overlook sin; his justice demanded that the account be settled.

Study Guide Questions on Chapter 3

1. Who is the serpent in verse 1? Is he another beast of the field?

2. Why do you suppose that Satan approached Eve instead of Adam?

3. Compare the dialogue between Satan and Eve in Genesis 3:1 – 4 with the instruction that God gives Adam in Genesis 2:16 & 17. List the ways in which the truth was altered in Chapter 3.

4. In verses 1 – 5 we find five common attacks that men make against the Word of God. List them.

5. There is a Bible Doctrine of Peccability – a five step process by which temptation turns to sin. Name (list) each of the five steps.

6. In verse 6 we see that the temptation of Eve was threefold and corresponds to the three part makeup of man. In other words, there are three areas in which man is subject to temptation. List them (Note John talks about them in 1John 2:15 – 16).

7. In verse 7 Adam and Eve learn for the first time that they were naked. Why did they not realize that they were naked before that?

8. Who were the gods that verse 5 talks about?

9. In what way is the apron of fig leaves representative of religion and religious works? Why was this covering not acceptable to God?

10. Why did man hide from God after they sinned? Do lost men do that today?

11. What is this thing called conscience? Conscience can compel us to do one of two different things with regard to God. What are they?

12. God's question to Adam "Where art thou?" proves two things regarding God in His interest in man. What are they?

13. List seven things that Adam lost when he fell into sin. Who did he blame for the loss?

14. Verse 15 is often called the proto-evangel – the first gospel. In that verse there are four things that would be accomplished by Christ. List them. Could man do anything to help himself in any of these things or does this all have to be God's work of redemption?

15. Can you identify who the seed of the woman is in verse 15? Can you identify the seed of the serpent?

16. What curse was put on the serpent in verse 14?

17. What is the ultimate outcome of the conflict between the seed of the woman and the seed of the serpent? Who finally wins the conflict?

18. List five results of the curse on the woman in verse 16.

19. List six results of the curse on the man in verses 17 – 19.

20. Do you see Adam's statement of faith in verse 20? How did he express it?

21. Verse 21 involves an act of God that results in the first blood that was shed in God's creation. What verse in Romans does this act ultimately point to?

22. Why were Adam and Eve barred from the tree of life in verses 22 through 24? What would have happened if Adam and Eve would have eaten of that tree?

23. The Dispensation of Innocence ended with man's sin in Genesis 3:6-7. The Dispensation of Promise starts in Genesis 3:15. Discuss how Romans 3:25; 5:12 and Galatians 3:19 relate to the Dispensation of Promise.

24. Genesis 3:16 implies that there is to be headship of the husband. Is this a natural tendency or is it something that requires diligence on the part of both the husband and wife?

25. Show from scripture how the curses put upon the man were all also born by Jesus Christ our Savior.

26. There are seven promises of the coming Savior in the Old Testament scriptures. List them.

## Genesis 4:1-7 (KJV)

[1] And Adam knew Eve[a] his wife; and she conceived, and bare Cain[b], and said, I have gotten a man from the LORD. [2] And she again bare his brother Abel. And Abel[c] was a keeper of sheep, but Cain was a tiller of the ground. [3] And in process[d] of time it came to pass, that Cain brought[e] of the fruit of the ground an offering unto the LORD. [4] And Abel, he also brought of the firstlings of his flock and of the fat thereof. And the LORD had respect unto Abel and to his offering: [5] But unto Cain and to his offering he had not respect[f]. And Cain was very wroth, and his countenance fell. [6] And the LORD said unto Cain, Why art thou wroth? and why is thy countenance[h] fallen? [7] If thou doest well, shalt thou not be accepted? and if thou doest not well, sin lieth at the [g]door. And unto thee *shall be* his desire[i], and thou shalt rule over him.

### [a] Adam knew his wife Eve

Eve is mentioned 4 times in the Bible: Genesis 3:20; 4:1; 2Corinthians 11:3; and 1Timothy 2:13. The term "knew his wife…" is to know her sexually so as to procreate as was God's design for man. In creating man, God created a race of free, moral, agents which could reproduce more of the same free, moral, agents. The race could reproduce itself and could do so endlessly. In Hebrews 2:6-8, we see that God had an eternal purpose for man – that purpose being the eternal custodianship of God's creation.

### [b] Birth of Cain

Eve names her first son "Cain" which means "gotten." She mistakenly believes that this is the promised seed of Genesis 3:15. However, it soon becomes apparent to her that Cain was "…of that wicked one" (1John 3:12). There is a father-child relationship between the natural man (1Corinthians 2:14) and the devil (John 8:44). Apart from personal faith (that being coming to God on God's terms) with the proper sacrifice, man remains what he is by nature -- lost and separated from God. Seeing the unbelief in her eldest son, Eve in frustration names her next son "Abel" which means "vanity." She did not understand the principle of 1Corinthians 15:46, "That was not first which was spiritual but that which was natural." This principle repeats itself over many times in scripture: The first is rejected as the wrong one.

| | |
|---|---|
| Two sons of Adam: | Cain – rejected for lack of faith |
| | Abel – accepted |
| Two sons of Abraham: | Ishmael – rejected for lack of faith |
| | Isaac – accepted |
| Two sons of Isaac: | Esau – rejected for lack of faith |
| | Jacob – accepted |
| Two kings of Israel: | Saul – rejected |
| | David – accepted |
| Two cities | Satan's city (Babel) came first – Genesis 11 |
| | Then God's – John 14 |
| Two Natures | The Old Nature came first |
| | Then the New |
| Two Messiahs | First the antichrist |
| | Then Christ |
| "Ye must be born <u>again</u>" | First birth (physical) |
| | Second birth (spiritual) |

## [c] Cain and Abel

Comparison of Cain and Abel: Both inherited Adam's sin nature. The difference between Cain and Abel was their sacrifice. The issue is not the person of the one doing the offering, but the character of the offering. Note [e] below will address that further.

However, there is also an interesting comparison between Adam and Christ. Adam, as father, was in a state of ruin and so were his children. "As is the earthy, such are they that are earthy" (1Cor. 15:45). Having inherited Adam's fallen nature, no member of Adam's race has the capability of working his way back to God. There is a Bible doctrine of Federal headship presented in Romans 5:12-21: Let's consider what that Federal headship entails. Before a person is saved today by faith in the shed blood of Christ, that person is "in Adam." Once saved, that person is trans-located and is "in Christ" and his or her sins are covered by the blood that the Savior shed on Calvary. Consider the following comparison of the headship of Christ and that of Adam as presented Romans 5.

[d]

### Table 7 Christ the Last Man and Adam Compared

| Adam | Christ |
|---|---|
| The First Man | The Last Man |
| By him Sin Entered | By Him Righteousness Entered |
| Disobedience | Obedience |
| Brought Death | Brought Life |
| Through his offence many be dead | Through him, the gift by grace abounded to many unto life |
| A Sin nature is derived from physical birth | New nature derived from regeneration by faith in Christ's work of redemption |
| Begotten by the will of man (John 1:13) | Begotten by God (James 1:18) |

## Cain's Sacrifice

"In the process of time…" means "at a proper time". There apparently was more information revealed to men at that time than is recorded for us in our Bible. The Bible gives us today what we need to know to come to God and to be fruitful. It does not contain every word God spake, for if it did "…even the world itself could not contain the books that should be written." (John 21:25). However, the presence of the Cherubim suggests the presence of God for He "…dwelleth between the cherubim…" (1Sam. 4:4; 2Sam. 6:2; 1Chron. 13:6) In every other place in Scripture, the cherubim indicated the presence of God. We assume this to be the case here as well. Genesis 4:7 speaks of a door to a tabernacle. This tabernacle could have been at the east gate of Eden. They knew that they were to bring a sacrifice to a specific place. They also knew what they were to bring. Hebrews tells us this, "by faith Abel brought a more excellent sacrifice than Cain." And "Faith cometh by hearing…by the Word of God" (Rom. 10:17).

## [e] Their Sacrifices Compared

We consider now the sacrifices of each of Cain and Abel. Just as Adam had to look outside of himself for redemption and the power to stand accepted by God, so his children each had the same decision to make.

Cain:   Brought the fruit of the cursed ground.

There was no blood to remove the stain of sin.

There was no presentation of a sacrificed life.

He tried to bring the LORD down to his level.

He operated on human wisdom – what better sacrifice than one's own labor?

But faith seeks God's wisdom and comes to God in and by God's way.

God teaches and faith listens.

His sacrifice demonstrated ignorance of:

1) his own condition and

2) God's holy character. God is not worshiped with man's hands (Acts 7:48; 17:24, 25).

He tried to make God the receiver, not the giver.

"It is more blessed to give than to receive" (Acts 20:35).

"And without all contradiction, the less is blessed by the better" (Heb. 7:7).

"Who hath first given to him?" (Rom. 11:35)

Abel:   Brought the "blood" (the symbol of a surrendered life) and the best of the flock -- the "fatness."

The Blood -- The relinquishing of innocent life speaks of the sacrifice of Christ.

The Fatness -- The best is offered. This speaks of the excellence of Christ.

Both were forbidden to be eaten under the Law of Moses. That is because the Law could not produce life. (Gal. 3:21).
But -- "Except ye eat the flesh of the Son of Man…ye have not life" (John 6:53). "The life of the flesh is in the blood…upon the altar" (Lev. 17:11)
Our Lord's resurrection body had flesh and bone (Luke 24:39) without reference to blood.

## [f] Results of Their Sacrifices

Their sacrifices produced different results:
> Abel's offering was brought in faith and filled his heart with peace.

Faith Justifies (Rom. 5:1), Purifies (Acts 15:9), Works by Love (Gal. 5:6), and Overcomes the World (1John 5:24)
> Cain's offering was brought in self will and filled his heart with pride and wrath.
> Unbelief -- He was too refined to bring a blood sacrifice.

But He was not too refined to shed innocent blood.
"But as then, he that was born after the flesh persecuted him that was born after the spirit." (Gal. 4:29)
> Cain came his own way - "The way of Cain" (Jude 12).

He brought the works of his own hands, the fruit of a cursed ground. Religion is cursed (Mark11:21).
Abel's sacrifice was "accepted" by fire (Leviticus 9:24 and Judge 6:21)
> 1Kings 18:38 (Elisha on Mt. Karmel)
> 1Chronicles 21:26

Why did God use fire to accept the sacrifice? Because God's judgment on sin is thereby represented. The fire of judgment consumes the substitute.
Psalm 20:3 "Remember all thy [God's] offerings and accept [lit. "turn to ashes" in the Hebrew] the burnt sacrifice.

## [g] The Place of the Sacrifice

Cain and Abel apparently had a mercy seat that they could come to.
- It was there on the earth up until the time of the flood. It was likely there at the East gate of Eden.
- After the flood, when Noah came off the Ark, we find for the first time, a burnt offering to God. This speaks of the mercy seat being taken away from the earth to heaven. This removing of the mercy seat apparently occurred just before the flood.
- The mercy seat was on earth again under the Law of Moses

## [h] Man's Access to God

Man, Man's Sacrifice, and the Presence of God
> <u>Man was in Innocence in Eden</u>
> - There he walked with God directly
> - He talked with God face-to-face
> - Apparently he could also see angels (the gods of Gen. 3:5).
> <u>Man was under the Dispensation of Promise after the fall but before the Flood</u>
> - He could talk with God but not face-to-face.
> - He had a specified place to offer sacrifices (at the East Gate of Eden).
> - Man had a specified sacrifice (a blood sacrifice).
> - Man had a specified time to offer the sacrifice.

- Man apparently had a prescribed place where the sacrifice was to be offered.
- The acceptable sacrifice was consumed by God with Fire.

Man after the Flood

- Had to build an altar because God had removed Himself and His place for the sacrifice from the earth.
- He had to offer burnt offerings so they could ascend up to the mercy seat in heaven.
- Man had a prescribed offering.
- 

Man under Law

- Man again had a mercy seat where God met with man.
    The mercy seat left earth when the captivity started.
    It was re-established under Ezra.
- Man again had a prescribed sacrifice.
- Man now had a prescribed priesthood.
- Man also had a prescribed form of service.
- With the giving of the Law to Israel, God re-established a mercy seat on the earth. This is seen in Exodus 25:8-22
    o Exodus 25:9 speaks of the Tabernacle.
    o Verse 10 speaks of the Ark of the Covenant.
    o Verse 17 addresses the mercy seat.
    o Verse 18. speaks of the Cherubim.
    o Verse 22 indicates that God met men there at the mercy seat.
- The mercy seat left again when Israel went into apostasy and was taken captive to Babylon.
- It was re-established again with Ezra.
- It was gone forever when Christ became the mercy seat (Rom. 3:25; 1 John. 2:2).

Man under Grace

- Believers are baptized into Christ upon believing that Christ paid sin's debt (Rom. 6:1-4).
- Believers are under a special and unique relationship of Christ to the Church.
- Christ the head together with the Church which is His body forms the One New Man (Eph. 2:15).

Man under the New Covenant (Israel's program)

- Christ will be the mercy seat.
- They will be joined to Christ by water baptism (Matt. 28:20; Mark 16:16; Acts 3:28).


## [i] The Lord Pleads with Cain

In verses 6 and 7, we find an interesting plea from the LORD to Cain. "Why art thou wroth? And why is thy countenance fallen? If thou doest well, shalt thou not be accepted?" These are rhetorical questions that the LORD asks Cain to get him to think about how gracious the LORD is. Let's consider each question.

1.  "Why art thou wroth?" What are you angry about Cain? Let's think about this and see if you have valid reason to harbor your anger and bitterness.
2.  "Why is your countenance fallen?" You can't hide the fact that you carry anger and bitterness. Your face tells everyone the attitude of your heart.
3.  "If thou doest well, shalt thou not be accepted?" This is a statement similar to Romans 2:6-9. There (in the Romans passage) we see "the righteous judgment of God" stated: God will "...render to every man according to his deeds."

- "To them who by patient continuance in well doing seek for glory and honor and immortality, eternal life." The system that was in effect then was that those people who, by perfectly, without fail, seek for glory and honor and immortality, would receive eternal life from God. Thus the LORD tells Cain that, if he does the same and did not sin, he would be accepted.
  - But, as Romans 3:10-12 states, "There is none righteous, no, not one. There is none that understandeth, there is none that seeketh after God. They are all gone out of the way, they are together become unprofitable, there is none that doeth good, no, not one." Again in Romans 3:22, 23 "…for there is no difference: For all have sinned, and come short of the glory of God."
4. The LORD continues saying "…and if thou doest not well, sin lieth at the door. And unto thee shall be his desire, and thou shalt rule over him." We can understand this if we consider that the Hebrew word for "sin" and "sin offering" are the same. God is telling Cain that a sin offering (probably a lamb) is lying at the door of the tabernacle. "…Unto thee shall be his desire;" He is there to be a willing sacrifice for your sins. "And thou shalt rule over him;" -- you can go over there and pick him up and he will not even run away from you. This sin offering is a type of Christ. He is "the lamb of God who taketh away the sin of the world" (John 1:29) "…Like a lamb dumb before his shearer, so opened he not his mouth" (Acts 8:32) "…In the midst of the elders stood a lamb as it had been slain," (Rev. 5:6) "…Worthy is the Lamb that was slain to receive power, and riches, and wisdom, and strength, and honor, and glory, and blessing" (Rev. 5:12) However, Cain was apparently too set in his determination that his offering should be sufficient to take advantage of the gift on the sin offering.

## Genesis 4:8-12 Cain slays his brother

[8] And Cain talked with Abel his brother: and it came to pass, when they were in the field, that Cain rose up[j] against Abel his brother, and slew[k] him. [9] And the LORD said unto Cain, Where[l] *is* Abel thy brother? And he said, I know[m] not: *Am* I my brother's keeper? [10] And he said, What hast thou done? the voice of thy brother's blood crieth unto me from the [n]ground. [11] And now *art* thou cursed from the earth, which hath opened her mouth to receive thy brother's blood from thy hand; [12] When thou tillest the ground, it shall not henceforth[o] yield unto thee her strength; a fugitive and a vagabond shalt thou be in the earth.

## [j] Cain's Rebellious Heart

In Genesis 4:8 Cain and Abel discuss the matter of the sacrifice. Likely, Abel brings up the subject and pleads with Cain to bring the appropriate sacrifice. Here we learn that Cain's bringing of the wrong sacrifice is not an innocent mistake. Had it been so, Cain would have humbly asked God "Then what do you want?" Instead, Cain responds in anger.

## [k] Different Results from Different Sacrifices

Aside from the animals that died to cloth Adam and Eve, the first animal to die was a sheep (Gen. 3:21). The first person to die was a shepherd. Both were types of Christ; the Good Shepherd who lays down His life for His sheep (Heb. 13:20).

Religion (as represented here by Cain and his offering) has possibly shed more blood than anything else on earth. God's way of salvation fills the heart with love. Man's way fills it with hatred.

## [l] The Improper Response to Conviction

God asks two questions:
1. "Where is Abel thy brother?" Compare this with the words "Where art thou?" of Genesis 3:9. This question is asked to convict. It speaks to the indwelling sin nature.
2. "What hast thou done?" This question is asked to bring a confession of personal sin and guilt.

The proper response to conviction is to acknowledge sin, avail oneself of God's mercy, and change the heart.

## [m] Cain's Lie

"I know not" is the response of an agnostic. Just as Cain lied, so does the modern day Agnostic. Ask an agnostic "Do you believe in God?" He responds, "I don't know." But he lies. Psalm 14:1 and Romans 1:20 both tell us that they do know that there is a God. They are just lying to themselves to enable themselves to feel good or somehow be justified in their unbelief.

Liberals of today say he is his brothers keeper and by this proceeds to enact legislation to control his brother. Cain is saying it is not my duty to take care of my brother, but it is ok for me to murder him.

## [n] No Place to Hide

Cain buried the evidence of his deed but finds that he cannot hide it from God. God knows of all of the sin committed in secret. The wise man reconciles himself to that fact.

## [o] Cain's Punishment

God brings a two-fold punishment on Cain:
1. He brought the works of his hand in tilling the ground; he would now no longer be blessed in his labor.
2. He did not have the courage to face his own guilt and come to God for redemption. Now his life would be one of fear. He is now a fugitive and his life is one in which he would not have rest and peace. He is a vagabond in the earth.

### Genesis 4:13-15 Cain's Punishment

13 And Cain said unto the LORD, My punishment[p] *is* greater than I can [q]bear. 14 Behold, thou hast driven me out this day from the face of the earth; and from thy face shall I be hid; and I shall be a fugitive and a vagabond in the earth; and it shall come to pass, *that* every one that findeth me shall slay me. 15 And the LORD said unto him, Therefore whosoever slayeth Cain, vengeance shall be taken on him [r]sevenfold. And the LORD set a mark upon Cain, lest any finding him should kill him.

## [p] Cain's Heart and David's

Cain's attitude toward the punishment meted out to him is not remorse for what he had done but sorrow for the severity of the punishment. Contrast this with David's remorse when his guilt was brought to his attention (Psalm 51:3) "For I acknowledge my transgressions: and my sin is ever before me. Against thee only have I sinned, and done this evil in thy sight…"

## [q] The Mark on Cain

"Everyone that findeth me…" speaks of a large population. 130 years could have produced a population of about 500,000 if the first child is born at age 21 and one / year thereafter. According to Genesis 5:3 and 4; Adam and Eve had many sons and daughters.

## [r] No Capitol Punishment for Cain

Capital punishment was not a part of God's plan for man at that point in time. God changed this after the flood (Gen. 9:5) when God said, "Whoso sheddeth man's blood, by man shall his blood be shed: for in the image of God made he man." (Gen. 9:6) Here, however, God reserves all vengeance and judgment to Himself.

## Genesis 4:16-18 Cain's Posterity

> [16] And Cain went out from the presence[s] of the LORD, and dwelt in the land of Nod, on the east[t] of Eden. [17] And Cain knew his wife; and she [u]conceived, and bare Enoch: and he builded a [v]city, and called the name of the city, after the name of his son, Enoch[w]. [18] And unto Enoch[x] was born Irad: and Irad begat Mehujael: and Mehujael begat Methusael: and Methusael begat Lamech.

## [s] Cain Turns His Back on the Lord

Cain went out from the presence of the LORD. That is, he left the tabernacle or the place where the sacrifice was to be made. He decided that he would no longer avail himself of the place of access to God.

## [t] Cain's Country

Cain moves east of Eden into the area that is today Iran, Iraq, India, and Pakistan.

## [u] Cain's Wife

Cain had a wife and took her with him. Before the Law, God allowed marriage between close relatives and even siblings. Adam's and Eve's children would have had to get their spouses from among their sibling. This would be no problem as it would be today in that the race was genetically purer then than it is today. Genetic mutation was apparently not an issue for the human race at that time. However, under the Law of Moses, God forbad marriage between close relatives. Neither genetic mutation nor somatic mutation was the problem that they are today. Somatic mutation weakens the individual and causes aging and eventually death. Genetic mutation weakens the species and that resulted in the forbidding of people to marry close relatives. The protection from mutations in the pre-flood world was likely due to a healthier environment.

## [v] Cain's City

Cain finds a large population in the area he traveled to and built a city. This is the first city built by man. He builds this city in rebellion to the curse put upon him—that he would be a vagabond in the earth. His city building activity is apparently a part of his scheme to thwart the punishment that God put upon him.

## [w] Cain's Cultured Descendants

In verse 17 we see that he begets a son in his image and names the city he built after his son (i.e. his own image). He proceeds to decorate the city with culture and probably with his religion as well. He and his descendants developed a sophisticated culture:

    Agriculture (vs. 20)

Music (vs. 21)
Metallurgy (vs. 22)
But also sin and violence:
Lust (vs. 19)
Pride and Arrogance (vs. 23): Note Lamech's disrespect for authority.
Violation of the Marriage of Institution (Matt. 19:3; 4:19-23)
Violence (vs. 23)
Contempt and Disdain for God (vs. 24)
Denial of Future Punishment (vs. 24)

## [x] Cain's Dedicated Son Enoch

In verse 17 we see the city dedicated to Enoch. There are two Enochs in Genesis (cf. 5:18). The name means "dedicated." Enoch, the son of Cain was dedicated to undoing the curse of Genesis 3:17 and 4:11. Cain is dedicated to making the world a better place, but doing so without God. He sets out to decorate the world with art, architecture and culture. He intends to make the world a respectable place and himself a respectable citizen of it. What Jude calls "the way of Cain" (Jude 11) was to reject God's effort to cleanse the world of sin and instead to supplant it with man's efforts to "improve" the world. The way of Cain produces by its inventive genius, all that the human mind can conceive of and devise. This is the world that we have around us today. What devices and what pleasures the eye sees the heart desires and these supplant the real need people have – the need of a Savior. Cain and his posterity built a beautiful culture on the ground that had been stained by the blood of Abel. Our society today builds its beautiful culture on a ground that is stained with the blood of Christ; but it goes on without giving it a thought. That blood, however, is what justifies the believer.

## Genesis 4:19-24 The Seed Line of Cain

[19] And Lamech took unto him two wives: the name of the one *was* Adah, and the name of the other Zillah. [20] And Adah bare Jabal: he was the father of such as dwell in tents, and *of such as have* cattle. [21] And his brother's name *was* Jubal: he was the father of all such as handle the harp and organ. [22] And Zillah, she also bare Tubalcain, an instructer of every artificer in brass and iron: and the sister of Tubalcain *was* Naamah. [23] And Lamech said unto his wives, Adah and Zillah, Hear my voice; ye wives of Lamech, hearken unto my speech: for I have slain a man to my wounding, and a young man to my hurt. [24] If Cain shall be avenged sevenfold, truly Lamech seventy and sevenfold.

Lamech is the great great grandson of Cain. There apparently was not much improvement in the spiritual state of this seed line.

## Genesis 4:25-26 The Godly Line of Seth

[25] And Adam knew his wife again; and she bare a son, and called his name [y]Seth: For God, *said she*, hath appointed me another seed instead of Abel, whom Cain slew. [26] And to Seth, to him also there was born a son; and he called his name Enos: then began men to call[z] upon the name of the LORD.

## [y] Birth of Seth

In verse 25, the narrative again picks up with Adam. Here, with the birth of Seth, the seed line is picked up again. Seth means "Substitute."

## [z] A Turn Back to the LORD

Seth begets Enos. With Enos it is said "…then began men to call upon the name of the LORD." Apparently, the way of Cain had been so all-pervasive that there was very little true worship until Seth and his descendants come along.

Genesis Chapter 4 Study Guide Questions

1. Based on Eve's words in verse 1, what would you consider Eve's thoughts are regarding Cain?

2. Why do you suppose he named the next son listed in the text as Abel meaning "vanity?"

3. 1Corinthians 15:46 states a principle "That was not first which was spiritual but that which was natural." List at least six examples from the Bible that bears this principle out.

4. Both Cain and Abel inherited the same sin nature from their father Adam. What was the difference that resulted in the divergent outcome of their spiritual states?

5. Romans 5:12 – 21 presents a Bible doctrine that we call "Federal headship." What does that say about every person who has been born into the human race?

6. Compare the headship of Adam with the headship of Christ.

7. Genesis 3:24 speaks of Cherubim being placed east of the Garden of Eden. What do you suppose they were placed there for? Cherubim in essentially every other place where we find them in scripture is associated with what?

8. It appears from this chapter that both Cain and Abel knew what they were supposed to bring as a sacrifice. How would they know that? Consider Genesis 3:21 in your answer.

9. There is one key difference between the two sacrifices. What was that key? There was an important key difference in the men that brought the sacrifices also. What was that difference? Consider Hebrews 11:4 and Romans 10:17 in your answer.

10. There is an element of human wisdom and pride in Cain's offering. Explain that.

11. In what way was Abel's sacrifice a type of Christ?

12. Compare the results of their sacrifices: Abel's sacrifice being brought in faith produced what in him? Cain's sacrifice being brought in self will filled him with what?

13. How did God demonstrate His acceptance of Abel's sacrifice? Consider Leviticus 9:24, Judges 6:21, 1Kings 18:38, in your answer. Why the fire?

14. Verse 7 speaks of a door. What does this suggest about the place of the offering of the sacrifice? Trace the mercy seat through history.

15. The relation of man, man's sacrifice, and God's presence changed through history. Describe how that relationship in each of the following time frames: 1) In Eden under innocence before the fall of man, 2) Under conscience before the flood, 3) After the flood, 4) When man was under the Law, 5) Today under grace, 6) When man will be under the New Covenant.

16. Answer the three questions that the Lord asked of Cain in verse 7...

17. What do you suppose was at the door of the tabernacle in verse 7? Would that be a type of Christ?

18. What does Cain's anger in verse 8 tell us about the condition of his heart? Was it just an innocent mistake on his part that he brought the wrong sacrifice?

19. Abel's sacrifice was a sheep. Abel himself was a shepherd. How are both of them a type of Christ?

20. There are pointed questions asked in Genesis 3:9 and in 4:8 and also in 4:10. They are intended for conviction. What is the proper response to conviction?

21. Cain's response to the question in verse 10 was the response of an agnostic "I don't know." The agnostic of today says he can't know God. What do the verses of Psalm 14:1 and Romans 1:20 say about that?

22. Verse 12 speaks of a twofold curse on Cain. What are they?

23. Verses 13 and 14 describe Cain's response to the punishment meted out on him. How does his response to the punishment compare with that of King David's response to his in Psalm 51:3?

24. How did the punishment for murder change from before the flood to after it (see Genesis 9:6)?

25. Verse 16 speaks of Cain going out from the presence of the Lord. We would understand that he is deciding that he will not avail himself anymore of what? Was that a smart move on his part?

26. Why was it acceptable for people to marry close relatives and even sibling before the Law of Moses?

27. Describe the culture and list the various skills that we see cultivated in Cain's descendents in verses 20 – 29.

28. Cain names his son Enoch (meaning dedicated). What was Cain dedicated to doing in building a city? Does this have anything to do with the curse that God put upon him for the murder of Cain? What is it that Jude calls the way of Cain in Jude 11?

29. What does the name Seth mean? What do you see starting to happen in the earth with the birth of Seth and of Seth's son Enos? Would this be a spiritual revival.

## The Generations of Adam

### Genesis 5:1-5 The Book of the Generation of Adam

[1] This *is* the book of the [a]generations[b] of Adam. In the day that God created man, in the likeness of God made he him; [2] Male and female created he them; and blessed them, and called their name Adam, in the day when they were [c]created. [3] And Adam lived an hundred and thirty years, and begat *a son* in his own likeness, after his image; and called his name Seth: [4] And the days of Adam after he had begotten Seth were eight hundred years: and he begat sons and daughters: [5] And all the days that Adam lived were nine hundred and thirty years: and he [d]died.

[a] The following are the names of the seed line of the patriarchs in consecutive order. This is one of those many instances that we encounter in the Bible to remind us that the author is not human but is God Himself. It also reminds us that God had the cross in mind from the beginning.  Names of the Seed Line:

Adam – Man
Seth – Appointed (Substitute)
Enos – A Mortal (Frail)
Cainan – Habitation (Dwelling Place)
Mahalaleel – The blessed God (The Praise of God)
Jared – Come down (Descend)
Enoch – Teaching
Methuselah – His death shall bring
Lamech – Captive
Noah – Comfort

Putting it all together:

Adam – Seth – Enos – Cainan – Mahalaleel – Jared
*Man appointed a mortal habitation (but) the Blessed God shall come down.*
Enoch – Methuselah – Lamech – Noah
*Teaching (that) his death shall bring the captives comfort.*

## [b] Nine subdivisions of Genesis

1. The generations of the heavens and the earth (Gen. 1:1-2:4)
2. The book of the generations of Adam (Gen. 2:4b-5:1)
3. The generations of Noah (Gen. 5:1b-6:9)
4. The generations of the Sons of Noah (Gen. 6:9b – 10:1)
5. The generations of Shem (Gen. 10:1b – 11:10)
6. The generations of Terah (Gen. 11:10b – 11:27)
7. The generations of Isaac (includes Ishmael) (Gen. 11:27b – 25:19)
8. The generations of Jacob (includes Esau) (Gen. 25:19b – 37:2)
9. The generations of the Sons of Jacob (Gen. 37: 2b – Exod. 1:1)

## [c] Origins in Genesis

Things CREATED by God
- The heaven and the earth (Gen. 1:1) A Space – Matter – Time Continuum
- Life (Gen. 1:21)
- Man (Gen. 1:27)

Things MADE by God
- The firmament (1:7)
- The sun and moon (1:16)
- The beasts of the earth (1:25)
- The garden of Eden (2:9)
- Eve was made from a rib from Adam (2:22)

Things INSTITUTED by God
- Volition – Free will
- Marriage
- Family
- Nationalism
- Atonement (this is not as clear as the above are but is none the less there as evidenced by the blood sacrifice of the animal that provided the skins that replaced the fig leaves of Genesis 3).

Things INTRODUCED by man
- Evil and spiritual death
- Religion
- Culture
    - Urbanization
    - Metallurgy
    - Music
    - Agriculture
    - Animal Husbandry
    - Writing
    - Education
    - Navigation
    - Textile
    - Ceramics

Things IMPOSED because of man's sin
- The curse on creation (2nd Law of Thermodynamics?)
- Languages (The confusion of Tongues at Babel)
- Death (both spiritual and physical)

**[d] Chapter 5 of Genesis is called the grave yard chapter of the Bible**.

Note the term, "and he died" is used eight times in the chapter.
(Adam –vs. 5, Seth – vs. 8, Enos – vs. 11, Cainan – vs. 14, Mahalaleel – vs. 17, Jared – vs. 20, Methuselah – vs. 27, Lamech – vs. 31).

Compare this with Matthew Chapter 1 where the emphasis is on life as the seed line is punctuated by "begetting."

> Abrahm begat Isaac
>> Isaac begat Jacob
>>> Jacob begat Judas
>>>> Judas begat Phares
>>>>> Phares begat Esram
>>>>>> Etc.

This difference in how the seed line is punctuated is because "…in Adam all die, even so in Christ shall all be made alive." (1Cor. 15:22)
This truth is stated again in Romans 5:12-14 "Wherefore, as by one man sin entered into the world, and death by sin; and so death passed upon all men, for that all have sinned. (For until the law sin was in the world; but sin is not imputed when there is no law). Nevertheless death reigned from Adam to Moses, even over them that had not sinned after the similitude of Adam's transgression, who is a figure of him that is to come."

## Genesis 5:6-21 Seth to Methuselah

[6] And Seth lived an hundred and five years, and begat[e] Enos: [7] And Seth lived after he begat Enos eight hundred[f] and seven years, and begat sons and daughters: [8] And all the days of Seth were nine hundred and twelve years: and he died.

[9] And Enos lived ninety years, and begat Cainan: [10] And Enos lived after he begat Cainan eight hundred and fifteen years, and begat sons and daughters: [11] And all the days of Enos were nine hundred and five years: and he died.

[12] And Cainan lived seventy years, and begat Mahalaleel: [13] And Cainan lived after he begat Mahalaleel eight hundred and forty years, and begat sons and daughters: [14] And all the days of Cainan were nine hundred and ten years: and he died.

[15] And Mahalaleel lived sixty and five years, and begat Jared: [16] And Mahalaleel lived after he begat Jared eight hundred and thirty years, and begat sons and daughters: [17] And all the days of Mahalaleel were eight hundred ninety and five years: and he died.

[18] And Jared lived an hundred sixty and two years, and he begat Enoch: [19] And Jared lived after he begat Enoch eight hundred years, and begat sons and daughters: [20] And all the days of Jared were nine hundred sixty and two years: and he died.

[21] And Enoch lived sixty and five years, and begat Methuselah:

## [e] The Bible actually provides a framework for the dating of the creation of Adam.

The basic information is in the Bible itself:
1. Genesis 5 contains chronological data from Adam to Noah.
2. Genesis 11 contains chronological data from Noah to Abraham.
3. Historical Books of the Old Testament (First and Second Samuel, First and Second Kings, First and Second Chronicles) contain chronological data from Abraham to Captivity.
4. Prophetic Books (Isaiah, Jeremiah, and Daniel) contain chronological data through the captivity to the rebuilding of Jerusalem under Ezra and Nehemiah.
5. The chronological data for the inter-testament period is presented clearly in the "seventy weeks" prophecy of Daniel Chapter 9.

Based on this data, the date for the creation of man is set at about 4000 B.C. See Appendix 1 for a compendium of various commentators who have studied Biblical chronology.

## [f] The question is often raised, "Did people really live for nearly a millennium?"

Up until the flood, the average life span of the first eight of the patriarchs was about 910 years.
There are several possible explanations for such longevity:
1. Atmospheric conditions were different. The concept of an antediluvian (pre-flood) water vapor canopy over the earth appears in many writings. 2Peter 3:15 also suggests such a canopy. In speaking of the days before Noah, Peter says, "For this they willingly are ignorant of, that by the Word of God the heavens were of old, and the earth standing out of the water and in the water: whereby the world that then was, being overflowed with water, perished:..." This speaks of water that is near the earth but not on the earth. It also speaks of a condition that was unique to the time before the flood. This water would be between the earth and the sun if it were truly effective in filtering ultraviolet radiation. This is not "...the waters which were above the firmament..." of which Genesis 1:7 speaks because the sun and moon exist in the firmament (open space) that was created by "the dividing of the waters from the

waters" (Gen. 1:14-17, cf. 1:6). Further Scriptural evidence for such a canopy is found in Genesis 2:5 where we find that in the antediluvian earth, "there was no rain upon the earth." Also, in Genesis 9:13, it is apparent that before the flood, there was no "bow in the clouds." Both conditions of which would be the result of such a water-vapor canopy.

Let's consider the effect that such a canopy would have:
   a. The earth would essentially be a greenhouse. The canopy would transmit incoming long wave radiation. Some of the reflected radiation would be retained, and some dispersed. The entire earth would have had an essentially uniform pleasant temperature over most of the earth.
   b. Uniform temperature would mean no high pressure and low pressure cells. Therefore, there would be no movement of great air masses and no violent storms.
   c. With no global air circulation, there would be no hydrologic cycle and no rain. Also, there would be no dust movement to the upper atmosphere and thus no means for precipitation of the canopy.
   d. There would be a uniform humidity maintained by a daily evaporation and condensation cycle.
   e. This global greenhouse would produce lush vegetation over essentially all of the earth. There would be few if any deserts or ice caps.
   f. The canopy would filter the destructive short wave radiation (ultraviolet radiation and cosmic rays) from the sun and from space. These short wave radiations produce mutations. Somatic (body) mutations weaken the individual and cause aging. Genetic mutation weakens the species. The water vapor canopy would therefore contribute to both human and animal health and longivity.
   g. The presence of the water vapor would contribute to (and tend to increase) atmospheric pressure. Research indicates that such hyperbaric pressures would actually be effective in combating disease and in promoting good health.

2.  There apparently were no diseases. Scripture does not mention diseases until God brought them upon Egypt (Exodus 15:26) when He brought Israel out of Egypt.

3.  There were no world wars; though there was severe violence.

## Genesis 5:22-32 Enoch and Methuselah

²² And Enoch walked with God after he begat Methuselah[g] three hundred years, and begat sons and daughters: ²³ And all the days of Enoch were three hundred sixty and five years: ²⁴ And Enoch walked with God: and he *was* not; for God took[h] him.
²⁵ And Methuselah lived an hundred eighty and seven years, and begat Lamech: ²⁶ And Methuselah lived after he begat Lamech seven hundred eighty and two years, and begat sons and daughters: ²⁷ And all the days of Methuselah were nine hundred sixty and nine years: and he died.
²⁸ And Lamech lived an hundred eighty and two years, and begat a son: ²⁹ And he called his name Noah, saying, This *same* shall comfort us concerning our work and toil of our hands, because of the ground which the LORD hath cursed. ³⁰ And Lamech lived after he begat Noah five hundred ninety and five years, and begat sons and daughters: ³¹ And all the days of Lamech were seven hundred seventy and seven years: and he died.
³² And Noah was five hundred years old: and Noah begat Shem, Ham, and Japheth.

## [g] Enoch walked with God "after" he begat Methuselah.

The name Methu-seleh is a compound word meaning "he shall go" plus "it shall come." Or literally "his death shall bring…"

Methuselah lived          187 years and begat Lamech

782 years that Methuselah lived after he begat Lamech

969 years of the life of Methusalah

Lamech lived     182 years and begat Noah

+187 years was Methuselah's age when he begot Noah

369 years = Methuselah's age when Noah born.

+600 (Noah's age at the Flood)

969 years from time Methuselah begotten to the flood.

Methuselah died the year of the flood (probably just before the flood).

Jude 14 tells us that Enoch prophesied judgment upon the ungodly.

Hebrews 11:5 says of Enoch, "By faith [he] was translated that he should not see death; and was not because God had translated him: for before his translation he had this testimony, that he pleased God."

Enoch was apparently given a revelation from God about the coming judgment of the flood and also about the delay in its coming. Enoch knew that as long as Methuselah was alive, the judgment would not come.

Methuselah's life span is the longest recorded of Genesis 5 (and the Bible).

Enoch's is the shortest of Genesis 5 (only 365 years—the number of days in a year).

There are two men in the Bible of whom it is said he would not see death, he pleased God, and he walked with God:

1. Would not see death ☐Enoch and John the Apostle
2. He pleased God ☐Enoch and Christ
3. He walked with God ☐Enoch and Noah

The flood happened 1656 years after Adam was created.

**[h] The Chronology of Genesis** can be determined from information given in Genesis:

Table 8: Chronology of Chapter 5

| AM | Man | Year | Event |
|---|---|---|---|
| 130 | Adam | @130 | Begat Seth |
| | | +800 | |
| | | =930 years | |
| 235 | Seth | @105 | Begat Enos |
| | | +807 | |
| | | = 912 years | |
| 325 | Enos | @90 | Begat Cainan |
| | | +815 | |
| | | =905 | |
| 395 | Cainon | @70 | Begat Mehalaleel |
| | | +840 | |
| | | =910 | |
| 460 | Mehalaleel | @65 | Begat Jared |
| | | +830 | |
| | | =895 | |
| 622 | Jared | @162 | Begat Enoch |
| | | +800 | |
| | | =962 | |
| 687 | Enoch | @65 | Begat Methuselah |
| | | +300 | |
| | | =365 | Translated in 987 AM |
| 874 | Methuselah | @187 | Begat Lamech |
| | | +782 | |
| | | =969 | (Died in 687+969=1656 AM) |
| 1056 | Lamech | @182 | Begat Noah |
| | | +595 | |
| | | =777 | Died in 777+874=1654 AM |
| 1556 | Noah | @500 | Begat Shem, Ham, Japheth |
| 1656 | Flood Happened | 1056+600 | Noah Begotten in 1056 Noah's recorded age at time of flood. |

AM for "Anno Munde" (Year of Man) means Years since creation of man

Genesis Chapter 5 Study Guide Questions

1. The names of the first ten patriarchs written in order of occurrence presents a message (makes a prophetic statement). Write that message out.

2. List the nine subdivisions of the Book of Genesis.

3. Make a list of each of the following things found in Genesis:

   - Things created by God
   - Things made by God
   - Things Instituted by God
   - Things Introduced by man
   - Things imposed because of man's sin

4. Why is Chapter 5 often called the graveyard chapter?

5. How would the atmospheric conditions of the pre-flood world have promoted long life spans?

6. What reason might God have for having man live longer in the early years of human history than later in our history? Consider passages such as Psalm 8:6-7 and Hebrews 2:6-9 in your explanation.

7. In what sense is Methuselah's name a prophecy? Is there spiritual significance to the fact that his is the longest life span recorded in the Bible? What does that say about God's attitude toward man?

8. Name two men in the Bible each of whom it was said:

   - "He would not see death…"
   - "He pleased God."
   - "He walked with God."

9. What does Jude 14 and Hebrews 11:5 have to say about Enoch?

# Genesis Annotations Chapter 6

## The Sons of God and the Daughters of Men

### Genesis 6:1-5 The Sons of God and the Daughters of Men

[1] And it came to pass, when men began to multiply on the face of the earth, and daughters were born unto them, [2] That the sons[a] of God saw the daughters of men that they *were* fair; and they took them wives of all which they chose. [3] And the LORD said, My spirit shall not always strive with man, for that he also *is* flesh[b]: yet his days shall be an hundred and twenty[c] years. [4] There were giants[d] in the earth in those days; and also after[e] that, when the sons of God came in unto the daughters of men, and they bare *children* to them, the same *became* mighty men which *were* of old, men of renown[f]. [5] And GOD saw that the wickedness of man *was* great in the earth, and *that* every imagination of the thoughts of his heart *was* only evil continually.

### [a] The Sons of God

Genesis 5:32 spoke about Noah and his sons: Genesis 6:8 speaks again of Noah. But the intervening verses digress to talk about the sons of God and the daughters of men. We ask the question here: "Who are the sons of God and who the daughters of men are?" There are two views on this:

1. The sons of God are the descendents of Seth (the seed-line), while the daughters of men are the daughters of the line of Cain.
2. The sons of God are angels while the daughters of men are just that—the daughters of the human race.

But believers in the Old Testament are not referred to as the "sons of God." They are in the New Testament however (John 1:12; Rom. 8:14, 19; Gal. 4:5; Phil. 2:15; 1 Jn. 3:1, 2). Adam is referred to as a "son of God" in Luke 3:38. Angels are referred to as "sons of God" (Job 1:6; 2:1; 38:1-7). Angels do not have wings -- the winged creatures of Zechariah 5 are female demons. Angels always appear as men (cf. Rev. 21:19) when they appear to men. They do not marry (Matt. 22:30) <u>in heaven</u> but there is no one for them to marry in heaven but on earth there is. Jude 6 and 2Peter 2:4 shed light on the question of who are the sons of God in Genesis 6. In Jude 6, we see a reference to "…the angels which kept not their first estate, but left their own habitation…" In 2Peter 2:4, we have a reference to "…the angels that sinned" in association with "…the old world that perished" in the days of Noah.

### "Sons of God" in Genesis 6:2

All free moral agents created directly by God are called "Sons of God."

1. Adam is called the son of God in the genealogy of Luke 3 (see Luke 3:38).
2. Those who are "in Christ" by regeneration are so called.
   a. They are called a new creature – 2Corinthians 5:17
   b. They have a new nature – Ephesians 2:10
   c. They are called sons of God in Romans 8:14,15
3. The believing remnant of Israel is called sons of God (John 1:13, 1John 3:1).
4. Angels are called sons of God in Job 1:6; 38:7; and Psalms 89:6.

Angels are also called "spirits" in Hebrews 1:14. Spirits are created by God (Psalm 104:4; Heb. 1:7, 14). Some of the angels fell into a special kind of sin (Jude 6). In Jude 6, they are said to have left their "habitation." In 2Corinthians 5:2, the same word is used of the resurrection body where it is translated "house." Jude 7 implies that Sodom and Gomorrah fell into the same sin. This activity of angels took place in the days of Noah (1Pet. 3:20; 2Pet. 2:7, 4). We ask how it could be that the sons of God (angels) could beget children by means of marriage with human women. One way might be by taking possession of the bodies of men. Such possession took place often during our Lord's earthly ministry (Matt. 8:16; 8:31; 9:33; 10:1, 8; 12:27, 28; 17:19; Mark 1:34; 5:9, 15; Luke 8:30). But then it was demons and not angels that took possession of the bodies of men. It might be also by direct marriage of angels with women in that angels, whenever they appear, appear as men as in the case of Gabriel in Daniel 9:21 or the angel in Revelation 21:17. Whatever the means, the progeny was genetically altered so as to be a corrupted seed line. This corrupted humanity had to be destroyed in order for "the seed of the woman" of Genesis 3:15 to be able to redeem man. God used the flood as the means of destroying the corrupted seed line. When this same eruption took place in the days of Abraham in the land of Canaan, Israel was to be the means of destroying them. This explains why certain of the people who occupied the land were to be not only driven out; but was to be totally destroyed. Deuteronomy 20:17 lists the people that were to be so destroyed (the Hittites, the Amorites, the Canaanites, and the Perizzites, the Hivites, and the Jebusites). We find this list again in Joshua 3:10. But Israel failed to drive all of them out (Josh. 13:13; 15:63; 16:10; 17:18; Judges 1:19-36).

## [b] He is Also Flesh

Verse 3 does not make sense unless we understand that the verse is talking about a corrupted humanity; one that is a mix of angelic (spirit) and human (flesh). The LORD is saying here that He will not always strive to salvage man in this corrupted form. Because this corrupted humanity is "also flesh," it can be totally destroyed by a flood.

## [c] 120 Years of the Longsuffering of the LORD

The 120 years here in verse 3 apparently is the time that is allotted to Noah for the building of the ark. This represents "the longsuffering of God" in the days of Noah (1Pet. 3:20.)

## [d] Giants in the Earth in Those Days

The giants are apparently the offspring of the conjugal relationship of the sons of God and the daughters of men. The Hebrew word for them is Ne-phil-im. It is translated "giants" here because they are great in size and in wickedness. The Hebrew word comes from the word "naphel" meaning to fall. They are superhuman and represent Satan's next effort to corrupt the seed line and thus to foil God's plan involving "the seed of the woman." (Gen. 3:15). The bringing of the flood upon "the world of the ungodly" (2Pet. 2:5) was necessary to destroy these superhuman creatures.

## [e] Also After That

There were giants in the earth "in those days" – in the days before the flood. But the verse goes on to say that there were giants in the earth again "after that." The giants are always the result of "...when the sons of God came in unto the daughters of men, and they bore children to them..." The production of the children was the apparent main reason for this activity, though it is also said they "saw the daughters of men that they were fair" (Gen. 6:2). This stresses the importance of women wearing modest apparel (1Tim. 2:9). We see the giants again in Numbers 13:33 "And there we saw the giants, the sons of Anak, which come of the giants: and we were in our own sight as grasshoppers; and so we were in their sight."

When God told Abraham that he would give him the land of Canaan (Gen. 11:31-12:1), Satan apparently proceeded to populate the land with his people. In Genesis 12:6, we see that "...the Canaanite was then in the land." In Genesis 15:17-21 we see God making the Abrahamic covenant with Abraham and giving him the land of Canaan which was occupied by the ten nations listed there. Among the ten nations were the Rephaims; a branch of the giants (Exod. 3:8, 17). They were to be completely destroyed (Deut. 20:17; Josh. 3:10). We see them referred to by other names as Emims (Deut. 2:10) and Zamzummims (Deut 2:20). Og, King of Bashan was one of the Anakim -- a family of the giants (Deut. 3:11). His bed was 4 cubits x 9 cubits (6' wide x 13.5' long). The giant Goliath and his brothers were likely some of these also.

## [f] Men of Renown

Because of their super natural origin and their great size, these giants become the leaders of society and renowned for wickedness.

## Genesis 6:6-7 God resolves to destroy the corrupted humanity

> [6] And it repented[g] the LORD that he had made man on the earth, and it grieved him at his heart. [7] And the LORD said, I will destroy man whom I have created from the face of the earth; both man, and beast, and the creeping thing, and the fowls of the air; for it repenteth me that I have made them.

## [g] The Lord Repented

In verse 6 we see that "it repented the LORD that he had made man on the earth." Man had changed and was no longer as God had created him. Therefore, God's attitude toward man changed.

## Genesis 6:8-10 But Noah found Grace

> [8] But Noah found grace[h] in the eyes of the LORD. [9] These *are* the generations of Noah: Noah was a just man *and* perfect in his generations[i], *and* Noah walked with God. [10] And Noah begat three sons, Shem, Ham, and Japheth.

## [h] Noah Found Grace

Noah found grace in the eyes of the Lord. This is not grace as in "the dispensation of grace" as we see it described in Ephesians 3:2. This is grace in a dispensation. God was gracious to save Noah and his family.

## [i] Noah Perfect in His Generations

Three things are said of Noah:
1. He was a just man. He did all that was required of him. It is interesting to note that the Old Testament saints are referred to as "just" while saints in the dispensation of grace are referred to as "justified."
2. He was "perfect in his generations." That means he was entirely human in his pedigree.
3. He walked with God. To walk with God is to say in effect -- he agreed with God (Amos 3:3).

## Genesis 6:11-12 A Corrupted Earth

<sup>11</sup> The earth[i] also was corrupt before God, and the earth was filled with violence. <sup>12</sup> And God looked upon the earth, and, behold, it was corrupt; for all flesh[k] had corrupted[l] his way upon the earth.

### [j] All the Earth Corrupt

In verses 11 and 12, we see that the "earth" was corrupt and "all flesh had corrupted itself upon the earth." This word translated corrupted is the same word translated "destroy" in verse 17. God would destroy his creation because it had been destroyed. To destroy something is to render it unfit for its intended use. We note that it says "all flesh had corrupted itself..." This corruption could have affected some of the animal creation as well as the human.

### [k] There are two clans of the Nephilim that can be identified in the Bible:

1. The descendents of Anak
2. The descendents of Repha

Anak:

Anak's father was Arba – the original builder of Hebron (Gen. 35:27 cf Josh. 15:13; 21:11). Anak had three sons who lived in Hebron (Judge 1:20). The descendants of Anak were called: the     Anakim (Numbers 13:33)

> The Emims (Deut. 2:10)
> The Avims (Deut. 2:23)
> The Zamzummims (Deut. 2:21)
> The Horim (Deut. 2:12, 22)

The Anakim were tall and powerful people.

> In Moab it was said of them "The Emims dwelt therein in times past, a people great, and many, and tall, as the Anakims; which also were accounted giants, as the Anakims; but the Moabites call the Emims. The Horims also dwelt in Sein before time; but the children of Esau succeeded them, when they had destroyed them from before them, and dwelt in their stead, as Israel did unto the land of his possession, which the LORD gave unto them." (Deut. 2:10-12).

> In Ammon the giants were: "The...Zamzummins: A people great, and many, and tall, as the Anakims..." (Deut. 2:21)

> "Hear, O Israel: Thou art to pass over Jordan this day, to go in to possess nations greater and mightier than thyself, cities great and fenced up to heaven, A people great and tall, the children of the Anakims, whom thou knowest..." (Deut. 9:1-2).

Repha:

The Rephaims were defeated by Chedorlaomer along with the Zuzims, the Emims, and the Horites (Gen. 14:5) in the campaign that included an attack on Sodom and Gomorrah. Comparing Genesis 15:21 with Exodus 3:8, 17 and 23:23, we see that they were mixed with the Canaanites, the Hittites, the Amorites, the Perizzites, the Hivites, and the Jebusites.

The word "giants" in Genesis 6:4 is from the Hebrew word "nephilim."

The word "giants" in Exodus 13:3 is from the Hebrew word "rephaim" as it is also in Numbers13:33 The word "rephaim" is transliterated 10 times "Rephaim":

- Genesis 14:5 "…Chedorlaomer…smote the Rephaim…"
- Genesis 15:20 "…Unto thy seed have I given this land…the Rephaims…"
- Joshua 12:15
- 2Samuel. 5:18 "The Philistines also come and spread themselves in the valley of Rephaim"
- 2Samuel 5:22 "…in the valley of Rephaim"
- 2Samuel 23:13 "… the Philistines pitched in the valley of Rephaim"
- 1Chron. 11:15 "…the Philistines camped in the valley of Rephaim"
- 1Chronicles 14:9 "And the Philistines came and spread themselves in the valley of Rephaim."
- 1 Chronicles 20:4 "And it came to pass after this, that there arose a war at Gezer with the Philistines; at which time Sibbechai the Hushathite slew Sippai, that was of the children of the giants [Repha].
- Isaiah 17:5 "the valley of Rephaim"

The Hebrew word "Rephaim" is also translated "dead" seven times:

Job 26:5 "<u>Dead</u> *things* are formed from under the waters"…
Psalm 88:10 "…shall the <u>dead</u> arise and praise thee?"
Proverbs 2:18 "For her house inclineth unto death, and her paths unto the <u>dead</u>."
Proverbs 9:18 "…But he knoweth not that the <u>dead</u> *are* there…"
Proverbs 21:16 "… shall remain in the congregation of the dead…"
Isaiah 14:9 "…it stirreth up the <u>dead</u> for thee…"
Isaiah 26:19 "…and the earth shall cast out the <u>dead</u>…"

It is translated "deceased" in Isaiah 26:14
It appears from Isaiah 26:14, 19 that they will have no resurrection. "[14] *They are* dead, they shall not live; *they are* deceased, they shall not rise: therefore hast thou visited and destroyed them, and made all their memory to perish."

## Genesis 6:13-22 God pronounces the destruction by the flood

[13] And God said unto Noah, The end[l] of all flesh is come before me; for the earth is filled with violence[m] through them; and, behold, I will destroy them with the earth.

## [l] Satan Attacks the Seed Line

What we are seeing in Genesis are attacks on the seed line by Satan. Such attacks were constant throughout the Old Testament. Before the creation of Adam, Satan attempted to thwart God's plan for the heavens by fomenting a rebellion there (Isa. 14:13 ff). With the creation of man, God declared that a creature that was created "lower than the angels" would be "…crowned with glory and honor…" and would "…have dominion over the works of [God's] hands." (Psalm 8:6-7; Heb. 2:6-9). Since the creation of man, Satan's attacks on God's purposes have been focused on the human race (2Cor. 4:4; Eph. 2:2). In each case, the human agent involved in the attack had his own personal interest to serve while Satan had his objective in view-- that being to prevent the fulfillment of the promise of Genesis 3:15. Let's consider some of those attacks in history and prophecy nn chronological order:

1. The temptation of Eve and the fall of Adam (The first gospel follows this first attack in the form of the promise of Gen. 3:15).
2. The murder of Abel (the apparent seed line) by Cain.

3. The development of a sophisticated culture to draw people off course to a godless civilization and religion that was devoid of the blood sacrifice.

4. The attempt to corrupt the human race by intermarriage of human women with angels (the sons of God of Genesis 6:2). According to Genesis 6:6, Noah and his family were pure in his generations Genesis 6:6.] The corrupted humanity, however, was destroyed by the flood.

5. The next attack was the introduction of idolatry with the Babylonian system. This was a threefold attack that involved a political system (one world government), an economic system (the state would provide for you so you do not have to depend on God), and a religious system (the worship of the hosts of heaven). This was all represented by the tower of Babel. Babel was built as a symbol of rebellion to God (Gen. 11:4). God countered this effort by confounding the languages of the people. At Babel God rejected the Gentile nations (Rom. 2:21-28).

6. With the rejection of the Gentile nations in Genesis 10, God separated Abraham as the channel of blessing to the Gentiles. As soon as it was revealed that the seed would come through Abraham, Satan again led an eruption of angelic interaction and intermarriage with human women. (Genesis 6:4, "and also after that.") In Genesis 12:6, we read that "the Canaanite was then [i.e. already] in the land." His aim was to occupy the land in advance of Abraham and contend for the right to it much as the Palestinian does today in the land of Israel.

7. Satan attempted to foil God's plan in Genesis 12:10-20 by the famine and Abraham's fear. We see this attempt repeated in Genesis 20:1-18.

8. The attempted destruction of Abraham's family by famine (Genesis 50:20).

9. The Egyptian captivity.

10. The attempted destruction of the male line at the time of the promised deliverance from Egypt (Exod. 1:10, 15; 2:5; Heb. 11:23).

11. The attempted destruction by Pharaoh in Exodus 14.

12. The attempted destruction of David (God's chosen king) by Saul (2Sam. 7).

13. The attempted destruction of David's seed line by Jehoram and Athaliah (2Chron. 17:1; 21:4).

14. The destruction of Jehoram's children except Ahaziah (2Chron. 21:17; 22:1)

15. The destruction of Ahaziah's children except Joash (2Chron. 22:10).

16. Hezekiah being childless (Isa. 36:1; 38:1; Psalm 136)

17. Haman's treacherous attempt to destroy the Jews (Esther 3:6, 12, 13; cf. 6:1)

18. Joseph's fear (Matt. 1:18-20 cf. Deut 24:1) was undone by a word from the angel.

19. Herod's jealousy (Matt. 2) resulted in the death of many children 2 years old and under.

## [m] A Tribute to Noah's parenting

The reason for the destruction of all flesh is that the earth was corrupt and filled with violence. The fact that Shem, Ham and Japheth had escaped the corruption is a testimony to Noah's and his wife's parenting skills. What probably also helped was the fact that they were busy building the ark for over a century. This in itself was also a testimony to parenting skills in that they were being asked by their dad to build a large barge in a world in which it had never rained. As we imagine the ridicule that the family was subjected to we can start to relate to them in our times when we speak of judgment to come.

## Genesis 6:14-16 God gives the plans for the Ark

<sup>14</sup> Make thee an ark of gopher wood; rooms shalt thou make in the ark, and shalt pitch it within and without with pitch. <sup>15</sup> And this *is the fashion*[n] which thou shalt make it *of*: The length[o] of the ark *shall be* three hundred cubits, the breadth of it fifty cubits, and the height of it thirty cubits. <sup>16</sup> A window[p] shalt thou make to the ark, and in a cubit shalt thou finish

it above; and the door of the ark shalt thou set in the side thereof; *with* lower, second, and third *stories* shalt thou make it.

## [n] The ark was a barge:

The ark was designed for capacity and stability and not for speed or navigation. It was more on the order of a barge than a ship. It was not designed to go anywhere but to float on the face of the universal flood which covered the earth.

## [o] The Dimensions of the ark:

Using a conversion unit of 1 cubit = 18", the dimensions of the ark would have been 450' long, 75' wide and 45' high. The ark was built with three stories in it. The total capacity of the ark would have been in excess of 1,500,000 cubic feet. This would be more than the capacity of 500 standard railroad box cars of our day.

## [p] The Window and the Door in the ark

The ark had an opening at the top 18" high that ran around the entire circumference. This would have been necessary for ventilation and for light. There is also one door that was to be placed in the side of the ark. This door is actually a type of Christ who is the door (John 10:7-9) and who is the only way to salvation. We will see more on this door later.

## Genesis 6:17-18 God will make a covenant with Noah

[17] And, behold, I, even I, do bring a flood[q] of waters upon the earth, to destroy all flesh, wherein *is* the breath of life, from under heaven; *and* every thing that *is* in the earth shall die. [18] But with thee will I establish my covenant[r]; and thou shalt come into the ark, thou, and thy sons, and thy wife, and thy sons' wives with thee.

## [q] The Uniqueness of the Flood

The word for "flood" used here is used only here and in Psalm 29:10. This was not a normal flood. It was a universal flood of truly "Biblical" proportions. It was intended to do exactly what it did – destroyed all flesh (except for that which was in the ark).

## [r] The First Reference to a Covenant

Here in verse 19 is the first reference to a covenant in the Bible. This covenant is made with the entire human race which is now confined to Noah, his wife, their sons and their sons' wives. The details of this covenant will be seen in Genesis 9:9-17.

## Genesis 6:19-22 Preservation of the animal creation

[19] And of every living thing of all flesh, two of every *sort* shalt thou bring into the ark, to keep *them* alive[s] with thee; they shall be male and female. [20] Of fowls after their kind, and of cattle after their kind, of every creeping thing of the earth after his kind, two of every *sort* shall come unto thee, to keep *them* alive. [21] And take thou unto thee of all food[t] that is eaten, and thou shalt gather *it* to thee; and it shall be for food for thee, and for them. [22] Thus did Noah; according to all that God commanded him, so did he.

## [s] A Genetic Choke Point:

We see here in the flood of Noah a genetic choke point for the gene pool for the living creatures on earth today. All of the life that we see and observe today is the descendency of the living creatures that were on the ark.

## [t] A one year food supply

Not only was the ark to house the animals but also the food that was to be eaten during the course of one year.

Study Guide Questions on Genesis Chapter 6

1. Who do you think the sons of God in verse 2 are? What are angels called in Job 1:6; 2:1; and 38:1-7? Are there any places in the Old Testament scriptures where men are called sons of God? Who are the daughters of men in Genesis 6:2? Are believers in the Old Testament ever referred to as "sons of God"? How about in the New Testament as in John 1:12 and in Romans 8:14 & 15.

2. Genesis 3:15 talks about the seed of the woman crushing the serpents head. Would it be advantageous to Satan if he could corrupt the human seed line that it was no longer a purely human seed line?

3. Who would be the angels that "kept not their first estate" in Jude 6? Who would be the angels that sinned…" that Peter speaks of in 2 Peter 2:4? Who would be the ungodly in 2 Peter 2:5?

4. What would be the significance of the words "…for that he is also flesh…" In verse 3?

5. There were giants in the earth associated with the intermarriage of the angels with human women. When would then be the time frame which the words "and also after that…" in verse 4 be referring to?

6. There are four families of giants mentioned in the Bible. Name the families that correspond to the Bible references: Exodus 3:8-17; Deuteronomy 2:10; 2:20; 2:23; and 3:11.

7. Why was Noah and his family selected out of all of humanity at that time to survive the flood? What was Noah's pedigree like?

8. Verse 12 says that "…all flesh had corrupted his way…" This would include man but what other kinds of flesh could be involved? Consider the reference to the "…strange flesh…" that Jude 1:7 speaks of.

9. According to verse 18, where was God when the flood was about to start? What then is the ark a type of in consideration of salvation?

## Noah and the Flood

### Genesis 7:1-4 Forty days and forty nights of rain

[1] And the LORD said unto Noah, Come[c] thou and all thy house into the ark; for thee have I seen righteous[d] before me in this generation. [2] Of every clean beast thou shalt take to thee by sevens, the male and his female: and of beasts that *are* not clean by two, the male and his female. [3] Of fowls also of the air by sevens, the male and the female; to keep seed alive upon the face of all the earth. [4] For yet seven days, and I will cause it to rain[a] upon the earth forty days and forty nights; and every living substance[e] that I have made will I destroy[f] from off the face of the earth.

### [a] Two questions come to mind regarding the flood of Noah in verse 4:

1. Where did the water come from to produce a flood of such Biblical proportions?
2. Where did the water go to after the flood?

Where did the water come from?

We understand from Genesis 2:5 that the climatic phenomenon before the flood was different than they are today. "...for the LORD God had not caused it to rain upon the earth, and there was not a man to till the ground." The rain did not come from the earth's atmosphere. We know from science that the earth's atmosphere can not hold enough water to cover the highest mountain by 15 cubits (about 25 feet according to Gen. 7:20). In Genesis 7:11, we see that the water came from two sources:

- The fountains of the great deep were broken up.
- The windows of heaven were opened.

The fountains of the great deep could refer to the points from which subterranean water went up as a mist to water the earth as we saw in Genesis 2:5. This would be somewhat as the geysers do today in Yellowstone Park and in parts of Russia. These subterranean features (whatever they were) no longer exist today because they were "broken up" during the onset of the flood. This would have to have involved the subsiding of the earth's crust in order to provide enough water to cover all of the land mass of the earth. It seems unlikely that this was the mode of the flooding of the earth.

The fountains of the great deep could also refer to the waters that were above the firmament (See note [b] below). Note: as we will see later in Chapter 11, the mountains that were there before the flood likely were not as high as they are now. We will study more on seismic and volcanic activity when we study about the physical dividing of the earth in the days of Peleg in Chapter 11.

The great deep could also be a reference to the water at the boundary of the universe – the boundary that separates the second heaven from the third heaven (2Cor. 12:1). If so, then "the deep" is the surface of that water and the deep has a firmament (or open space) in which the universe exists.

The windows of heaven are apparently openings that can be opened or closed in the top of the firmament that we learned of in Genesis 1:7, 8, and 14. The firmament (i.e. the open space) encompasses the sun, moon and stars (Gen. 1:14). There is water above that firmament and water below it. Both are called "seas" by God "And God said, Let the waters under the heaven be gathered together unto one place, and let the dry *land* appear: and it was so. And God called the dry *land* Earth; and the gathering together of the waters called he Seas: and God saw that *it was* good." (Gen. 1:9-10) This is the water that came through the windows of heaven. Appendix 20 gives a more thorough discussion of the source of the water that produced this world inundating flood. The "...windows of heaven..." would then be pathways through the stellar universe by which the waters that inundated the earth in Genesis 1:2 were removed from the earth to the outer boundary of the universe to form a part of the separation between the second heaven (the stellar universe) and the third heaven where God's throne is today.

**Where did the water go?**

In Genesis 8:1, we find that "...God made a wind to pass over the earth and the waters assuaged..." In Genesis 8:3, we see the results "And the waters returned from off the earth continually..." God returned the waters that were above the firmament back to the place from which it came to produce the flood. Some of the water that was in the fountains of the great deep is now in our oceans. Those fountains that were in the earth's crust (if indeed they existed) now no longer exist. Rather, the earth is now covered with sedimentary rock formations and soil deposits that contain most of our ground water today.

**[b] The Fountains of the Deep**

The "Deep" can be a reference to the oceans and seas but it can also be a reference to the "...waters that are above the firmament." in Genesis 1. Consider the following verses:

> [2] And the earth was without form, and void; and darkness *was* upon the face of the deep. And the Spirit of God moved upon the face of the waters. (Genesis 1:2)

This was the water that inundated the earth as we saw it in Genesis 1:2.

> [11] In the six hundredth year of Noah's life, in the second month, the seventeenth day of the month, the same day were all the fountains of the great deep broken up, and the windows of heaven were opened (Genesis 7:11)

This could be the water that produced the mist that watered the pre-flood world that Genesis 2:6 spoke of.

> [2] The fountains also of the deep and the windows of heaven were stopped, and the rain from heaven was restrained; (Genesis 8:2)

This is again a reference to the waters of Genesis 7:11.

> [13] And of Joseph he said, Blessed of the LORD *be* his land, for the precious things of heaven, for the dew, and for the deep that coucheth beneath, (Deut. 33:13)

This is apparently the water between the second and the third heaven.

> [30] The waters are hid as *with* a stone, and the face of the deep is frozen. (Job 38:30)

This would be the floor of the third heaven.

> [31] He (Leviathan) maketh the deep to boil like a pot: he maketh the sea like a pot of ointment. (Job 41:31)

This would be the water on the earth's surface today.

> [6] Thou coveredst it with the deep as *with* a garment: the waters stood above the mountains. (Psalm 104:6)

This could be either the waters of Genesis 1:2 or the water of Genesis Chapters 6 and 7.

> [23] They that go down to the sea in ships, that do business in great waters; [24] These see the works of the LORD, and his wonders in the deep. (Psalm 107:23-24)

This is a reference to the waters in the oceans and seas today.

[27] When he prepared the heavens, I *was* there: when he set a compass upon the face of the depth: [28] When he established the clouds above: when he strengthened the fountains of the deep: (Prov. 8:27-28)

This is a description of the hydrologic regime of the pre-flood world.

[27] That saith to the deep, Be dry, and I will dry up thy rivers: (Isa. 44:26-27)

[10] *Art* thou not it which hath dried the sea, the waters of the great deep; that hath made the depths of the sea a way for the ransomed to pass over? (Isa. 51:10)

These verses describe God's ability to regulate the rivers and seas as He did in the Exodus

With these verses in mind, let's revisit Genesis 1:7, 8, 9, 10, &14

"[7] And God made the firmament, and divided the waters which *were* under the firmament from the waters which *were* above the firmament: and it was so. [8] And God called the firmament Heaven. And the evening and the morning were the second day. [9] And God said, Let the waters under the heaven be gathered together unto one place, and let the dry *land* appear: and it was so. [10] And God called the dry *land* Earth; and the gathering together of the waters called he Seas: and God saw that *it was* good. [11] And God said, Let the earth bring forth grass, the herb yielding seed, *and* the fruit tree yielding fruit after his kind, whose seed *is* in itself, upon the earth: and it was so. [12] And the earth brought forth grass, *and* herb yielding seed after his kind, and the tree yielding fruit, whose seed *was* in itself, after his kind: and God saw that *it was* good. [13] And the evening and the morning were the third day."

[14] And God said, Let there be lights in the firmament of the heaven to divide the day from the night; and let them be for signs, and for seasons, and for days, and years: (Gen 1:7-14)

Note that the sun and the moon are in the firmament of the heaven. That means that the waters that are above the firmament are out beyond the sun and the moon. That water is out beyond the solar system and in fact is out beyond the universe. The term "seas" and "waters" in verse 10 are plural. There are two seas. There is a sea under the firmament (the open space of the universe) and a sea above that open space.

## [c] "Come into the Ark"

Genesis 7:1 The LORD tells Noah "Come thou …into the ark…" This tells us that God was in the ark. As we consider passages as 1Peter 3: 18-21 and 2Corinthians 5:19-21 we see that the ark was a type of Christ. 2Corinthians 5:19 "…God was in Christ reconciling the world unto himself…" But in Genesis 7 there was only one family that would respond to the invitation to "come." 1Peter 2:20 speaks of this time "…when the long suffering of God waited in the days of Noah, while the ark was a preparing, wherein few, that is eight souls were saved by water." The same water that destroyed the world that perished also bore up the ark to save those in it. In much the same manner, Christ bore the wrath of God against sin but only those who are in Christ by faith are saved from that wrath. According to 1Peter 2:20, water baptism placed Peter's hearers into Christ. However, in Paul's epistles it is the baptizing work of the Holy Spirit that places people into Christ (1Cor. 12:12 & 13; Rom. 6:1-4; Gal. 3:27; et. al). The Holy Spirit baptizes a person into a spiritual union with the Lord Jesus Christ so as to enable Christ to pay sin's debt on behalf of the believer.

The invitation "Come…" is repeated in Matthew 11:28 "Come unto me all ye that labor and are heavy laden, and I will give you rest…" and again in Revelation 22:17 [17] And the Spirit and the bride say, Come. And let him that heareth say, Come. And let him that is athirst come. And whosoever will, let him take the water of life freely."

### [d] The Righteousness of Faith

Note: "…for thee have I seen righteous before me in this generation." We ask: "Why was Noah seen as righteous?" The answer is in Hebrews 11:7 "By faith Noah …built and ark…" We should also note in verse 16 that the LORD "shut him in." Noah was safe and secure in the ark while the judgment of the flood was upon the world outside. Let's imagine though for a while what kind of scoffing Noah must have endured for 120 years while he and his sons were building an ark in a world where it had never rained (Gen. 2:5). There was a different atmosphere at that time. Rather than a hydrologic cycle as we have today, there was a vapor cycle involving evaporating and condensing of a vapor or mist. One can well imagine the heckling that would go on in such a case. The same scoffing will be endured by the tribulations saints when they speak of the judgment that will come on the world after the tribulation at the Lords return.

### 2Peter 3:3-10

"[3] Knowing this first, that there shall come in the last days scoffers, walking after their own lusts, [4] And saying, Where is the promise of his coming? for since the fathers fell asleep, all things continue as *they were* from the beginning of the creation. [5] For this they willingly are ignorant of, that by the word of God the heavens were of old, and the earth standing out of the water and in the water: [6] Whereby the world that then was, being overflowed with water, perished: [7] But the heavens and the earth, which are now, by the same word are kept in store, reserved unto fire against the day of judgment and perdition of ungodly men.

[8] But, beloved, be not ignorant of this one thing, that one day *is* with the Lord as a thousand years, and a thousand years as one day. [9] The Lord is not slack concerning his promise, as some men count slackness; but is longsuffering to us-ward, not willing that any should perish, but that all should come to repentance. [10] But the day of the Lord will come as a thief in the night; in the which the heavens shall pass away with a great noise, and the elements shall melt with fervent heat, the earth also and the works that are therein shall be burned up." (2Peter 3:3-10)

### [e] Genesis and Geology

This passage (2Peter 3:4) "And saying, Where is the promise of his coming? for since the fathers fell asleep, all things continue as *they were* from the beginning of the creation…" speaks of what we call an uniformitarian approach to geology. Those who reject the authority of Scripture in favor of the human philosophies of men (such as Darwin, Kierkegaard, Wellhausen, et.al.) hold that the earth today is a product of time and chance over multiplied millions of years of uniform processes. However, as we study Genesis, we see that the forces that have shaped the earth were actually very catastrophic processes brought on as a result of judgments on the rebellion of its inhabitants. The Lord is slow to anger and long suffering with men as He waits for them to repent but then acts quickly and decisively in judgment when the time is right in His plan and purposes for the ages.

What we call the geologic column testifies to the fact that there had been a series of catastrophic events that have shaped our earth. If one were to take various sizes of soil particles (i.e. sand silts, clays, rocks, gravels, etc and put them in a very large beaker full of water along with some living organisms. One would have a representative sample of the churned up mass of material that would have existed in the waters that covered the earth in the flood of Noah. Now shake that beaker up and allow it to settle and you would have a representative sample of our geologic column. You would find conglomerates on the bottom, then the sandstones above it, the silt sized particles above them forming the shales and siltstone, and the clay above that forming the limestone and dolomites. Also, the simpler forms of life would be denser and would be on the lower strata while the lighter and more complex forms of life would be higher in the strata. As marine organisms that secrete cementious material such as silica and calcium compounds are trapped in the deposits, the sediment hardens to form sedimentary rocks.

Also, as vegetative matter is trapped in the sediments, we would see the deposits of fossil fuels (coal and petroleum). The petroleum would be the result of animal material (both aquatic and terrestrial) while the coal would be the result of floating vegetation buried in successive layers in the receding waters of the flood. Since prior to the flood, the climate would have been conducive to the growth of lush vegetation, there would be a lot of floating material.

Picture a cove or a bay being acted upon by the successive tides carrying floating vegetation during the period of the receding flood waters. As the tide comes in, it brings a fresh supply of vegetative matter which it leaves behind as the tide recedes. The next tide brings in a deposit of sediment to cover the previously deposited organics and another supply of vegetative matter only to leave that again as it recedes. The results would be the coal deposits as we find them in vast amounts on our continent.

I refer you to the figures "Genesis and Geology" in the Appendix 5 on glacial geology. This material will be covered in later sections when we get into Chapter 11 of Genesis.

One simple convincing test that invalidates the uniformitarian theory is the formation of fossils. There are no fossils being formed in the world today by nature. If, for example, a leaf were to fall to the ground and be buried in sediment of a stream, it would decompose before it would be fossilized. However, the fossils (both animal and vegetative) that we find in sedimentary rocks are preserved in great detail. This could only have occurred if they had been buried in deposits deep enough to have excluded air from access to the fossil to prevent the decomposing to enable the fossilization process to go on.

## [f] The Destruction of all Life

God said "…every living substance that I have made will I destroy from off the face of the earth…" We need not look too far to find the evidence of this mass destruction of animal life. There are a number of references that provide information of the witness of the animal bone yards in many places on the earth where bone of many different species of animals (many of which have since gone extinct) are mingled together in one place. These animals had apparently tried to escape to higher ground until there was no higher ground to escape to anymore. In the book "The Genesis Flood" Henry Morris documents by photographs the footprints of a man and a dinosaur in the same rock strata and in fact in the same photo. It is noted also that the human foot print was unusually large. Remember: "There were giants in the earth in those days."

## Genesis 7:5-9 - The flood starts on the 17th day of the second month

> [5] And Noah did according unto all that the LORD commanded him. [6] And Noah *was* six hundred years old when the flood of waters was upon the earth. [7] And Noah went in, and his sons, and his wife, and his sons' wives with him, into the ark, because of the waters of the flood. [8] Of clean[g] beasts, and of beasts that *are* not clean, and of fowls, and of every thing that creepeth upon the earth, [9] There went in two and two unto Noah into the ark, the male and the female, as God had commanded Noah.

## [g] Clean and Unclean Animals

In Genesis 7:2 we see that the clean animals were brought in by sevens while those regarded as not clean were by twos. The seven would be divided into two and five. Five would be for sacrifices while the two for reproduction.

## Genesis 7:10-12 - The 600th year of Noah's life

[10] And it came to pass after seven days, that the waters of the flood were upon the earth. [11] In the six hundredth[h] year of Noah's life, in the second month, the seventeenth day of the month, the same day were all the fountains of the great deep broken up, and the windows of heaven were opened. [12] And the rain was upon the earth forty days and forty nights.

## [h] The Time of the Flood

The flood begins on the 600th year of Noah's life. If one were to add up the years using the chronology of the Book of Genesis, one would find the flood was apparently in the 1656th year after the creation of Adam The time line for the events of the flood are seen in chapter 8. The time line is as follows:

| | |
|---|---|
| Genesis 7:12 | 40 days of rain |
| Genesis 7:24 & 8:3 | 150 days from the start until the waters peaked |
| Genesis 8:5 | 220 days into the flood, the tops of the mountains were seen (They had been covered with at least 25' of water – 7:20) |
| Genesis 8:6-9 | 263 days into the flood the raven and dove were sent out |
| Genesis 8:10-11 | 270 days into the flood the dove was sent out again Returned with an olive branch |
| Genesis 8:12 | 277 days into the flood the dove was sent out but did not return |
| Genesis 8:13 | 313 days into the flood the waters dried up and the cover was opened |
| Genesis 8:14 | 374 days after the start of the flood God tells Noah to go forth out of the ark |

## Genesis 7:13-24 - The Flood

[13] In the selfsame day entered Noah, and Shem, and Ham, and Japheth, the sons of Noah, and Noah's wife, and the three wives of his sons with them, into the ark; [14] They, and every beast after his kind, and all the cattle after their kind, and every creeping thing that creepeth upon the earth after his kind, and every fowl after his kind, every bird of every sort. [15] And they went in unto Noah into the ark, two and two of all flesh, wherein *is* the breath of life. [16] And they that went in, went in male and female of all flesh, as God had commanded him: and the LORD shut[i] him in. [17] And the flood was forty[j] days upon the earth; and the waters increased, and bare up the ark, and it was lift up above the earth. [18] And the waters prevailed, and were increased greatly upon the earth; and the ark went upon the face of the waters. [19] And the waters prevailed exceedingly upon the earth; and all the high hills, that *were* under the whole heaven, were covered. [20] Fifteen cubits upward did the waters prevail; and the mountains were covered. [21] And all flesh died that moved upon the earth, both of fowl, and of cattle, and of beast, and of every creeping thing that creepeth upon the earth, and every man: [22] All in whose nostrils *was* the breath of life, of all that *was* in the dry *land*, died. [23] And every living substance was destroyed which was upon the face of the ground, both man, and cattle, and the creeping things, and the fowl of the heaven; and they were destroyed from the earth: and Noah only remained *alive*, and they that *were* with him in the ark. [24] And the waters prevailed upon the earth an hundred and fifty days.

## [i] In verse 16 we read "the Lord shut them in."

The Lord closed the door. This was after 120 years of God warning the pre-flood world through Noah of the judgment to come. This is similar to what will happen on earth after the rapture closes the dispensation of grace as we see in 2Thessalonians 2:11-12 "¹¹ And for this cause God shall send them strong delusion, that they should believe a lie: ¹² That they all might be damned who believed not the truth, but had pleasure in unrighteousness."

## [j] Types in Scripture
The number 40 here in verse 17 is the number for testing in the Bible. It reminds us the testing of Israel in the wilderness. There are some interesting types involved in the events of Genesis 6 thru 8.

> The number 40 is the number for testing and trial.
> Enoch - was translated without dying (type of the rapture?)
> The world that perished is a type of the world in the coming tribulation period.
> Noah – Preserved through the flood is a type of the tribulation saints.
> The Ark – Type of Christ and Water Baptism (1Peter 3:18ff)
> The Raven – Type of the Old sin nature
> The Dove – Type of the Holy Spirit
> The New World – Type of the Kingdom that will be set up on earth after the tribulation.

**God Remembers Noah**

## Genesis 8:1-22 (KJV) God Remembered Noah

[1] And God remembered[a] Noah, and every living thing, and all the cattle that *was* with him in the ark: and God made[b] a wind to pass over the earth, and the waters asswaged;

[2] The fountains also of the deep and the windows of heaven were stopped, and the rain from heaven was restrained;

[3] And the waters returned from off[c] the earth continually: and after the end of the hundred and fifty days the waters were abated.

[4] And the ark rested[d] in the seventh month, on the seventeenth day of the month, upon the mountains of Ararat.

[5] And the waters decreased continually until the tenth[e] month: in the tenth *month*, on the first *day* of the month, were the tops of the mountains seen.

[6] And it came to pass at the end of forty[f] days, that Noah opened the window of the ark which he had made:

[7] And he sent forth a raven[g], which went forth to and fro, until the waters were dried up from off the earth.

[8] Also he sent forth a dove[h] from him, to see if the waters were abated from off the face of the ground;

[9] But the dove found no rest for the sole of her foot, and she returned unto him into the ark, for the waters *were* on the face of the whole earth: then he put forth his hand, and took her, and pulled her in unto him into the ark.

[10] And he stayed yet other seven[i] days; and again he sent forth the dove out of the ark;

[11] And the dove came in to him in the evening; and, lo, in her mouth *was* an olive leaf pluckt off: so Noah knew that the waters were abated from off the earth.

[12] And he stayed yet other seven days; and sent forth the dove; which returned not again unto him any[j] more.

### Noah Opens the Arc

[13] And it came to pass in the six hundredth[k] and first year, in the first *month*, the first *day* of the month, the waters were dried up from off the earth: and Noah removed the covering of the ark, and looked, and, behold, the face of the ground was dry.

[14] And in the second month, on the seven and twentieth day of the month, was the earth dried.

[15] And God spake unto Noah, saying,

[16] Go forth[l] of the ark, thou, and thy wife, and thy sons, and thy sons' wives with thee.

[17] Bring[m] forth with thee every living thing that *is* with thee, of all flesh, *both* of fowl, and of cattle, and of every creeping thing that creepeth upon the earth; that they may breed abundantly in the earth, and be[m] fruitful, and multiply upon the earth.

[18] And Noah went forth, and his sons, and his wife, and his sons' wives with him:

[19] Every beast, every creeping thing, and every fowl, *and* whatsoever creepeth upon the earth, after their kinds, went forth out of the ark.

### Life Starts Over on Earth

[20] And Noah builded an altar[n] unto the LORD; and took of every clean beast, and of every clean fowl, and offered burnt offerings on the altar.

[21] And the LORD smelled a sweet savour; and the LORD said[o] in his heart, I will not again curse the ground any more for man's sake; for the imagination of man's heart *is* evil from his youth; neither will I again smite any more every thing living, as I have done.

[22] While the earth remaineth, seedtime and harvest, and cold and heat, and summer and winter, and day and night shall not cease.

**[a] God remembered Noah**. This is important for us to remember also. Sometimes we wonder if God really cares or if He has forgotten us.

**[b] Verse 1** God made a special wind that actually took the water off from the earth. This wind reversed the intake of water that came to earth from the waters that were above the firmament which came to earth through the windows of heaven at the start of the flood. This wind took the water off of the earth and took it back to the deep that was above the second heaven.

**[c]** In verse 3 we see the water that came from the windows of heaven (which we believe came from the water that was above the firmament) back to where it was before the flood.

**[d]** Here in verse 4 we see the ark resting on land in the seventh month in the 17th day of the month. This means only that the ark is no longer floating on the face of the waters. The waters had covered the highest mountains by 15 cubits. This would be about 25 feet. This tells us that the draft of the ark was at least that deep. Here the ark is caught on a high ground but the water still surrounds it. Remember in Genesis 7:11 it was in the second month in the seventeenth day of the month that the waters of the great deep were broken up.

**[e]** In verse 5 we see that on the first day of the tenth month the tops of the mountains were visible.

**[f]** After 40 more days Noah opened the windows that were built into the top story of the ark.

**[g]** In verse 7 we see him send out a raven but then in verse 8 he sends out a dove. The raven would find dead flesh to live off of but the dove would need some live plants.

**[h]** The dove's return told Noah that the water was not sufficiently abated to leave the ark.

**[i]** After seven more days he sends the dove again which returned with an olive leaf. The olive is a symbol of peace. That would have been reassuring for Noah.

**[j]** After another 7 days (verse 12) the dove was sent out again but did not return. That told Noah that there was now life on earth again.

**[k]** In verse 13 it was in the 601st year in the first day of the first month that the waters were dried from off the earth so that Noah could open the arc. However it was in the 27th day of the second month that the earth was dried enough to leave the arc

**[l]** In verse 16, God is still in the ark with Noah saying "Go forth out of the ark…" However in verse 17 He is outside of the arc bidding Noah and his family to bring forth every living thing out of the ark.

**[m]** Here in verse 17, God re-commissions man here to be fruitful and multiple upon the earth and he is again told to replenish the earth..

**[n] Noah offers a sacrifice as a sweet savor.**

This is the first mention of an altar in Scripture. Before the flood, there was a prescribed place at the east entrance to the Garden of Eden where man was to bring a sacrifice. That was where the cherub guarded the entrance (Gen. 3:24). That would have been where Cain and Abel brought their sacrifices. The East gate no longer existed on earth, having been destroyed by the flood. Therefore, an alter is now needed upon which the sacrifice would be burned.

**[o]** In verse 21 we find the LORD make a promise that He will never again flood the entire earth to destroy all flesh. However, one day the earth will face dissolution by fire as Peter says in 2Peter 3:3-7. However, that future event will not be a judgment per se but rather a remaking of the earth to go into the eternal state without any stain of sin.

**2Peter 3:3-**

3 Knowing this first, that there shall come in the last days scoffers, walking after their own lusts,

4 And saying, Where is the promise of his coming? for since the fathers fell asleep, all things continue as *they were* from the beginning of the creation.

5 For this they willingly are ignorant of, that by the word of God the heavens were of old, and the earth standing out of the water and in the water:

6 Whereby the world that then was, being overflowed with water, perished:

7 But the heavens and the earth, which are now, by the same word are kept in store, reserved unto fire against the day of judgment and perdition of ungodly men.

Study Guide Questions on Genesis Chapters 7 and 8

1. What two questions come to mind regarding the water that produced the flood of Genesis Chapter 7?

2. How would you answer the question "Where did the water come from?"

3. How would you explain how the water that covered even the highest mountains was removed from the earth and where did it all go when it was removed?"

4. Does Genesis 1:7-10 relate to the waters of Chapter 7?

5. In verse 1, God is speaking to Noah 7 days before the flood starts. According to that verse, where was God when He called to Noah?

6. On what basis did God declare Noah "Righteous" in verse 1? Romans 4:3 and Hebrews 11:7 can help you answer that question.

7. Why were seven of the clean animals taken into the ark while only two of each of the unclean were brought in?

8. How did the hydrologic cycle of the earth change from the time before the flood to after the flood was over? What happened during the flood event that would have caused that change?

9. 2Peter 3:3-10 speaks of men in the last days saying "…since the fathers fell asleep, all things continue as they were from the beginning of creation." This is a principle of history that we would call an uniformitarian view. Is this a valid concept from what we learn about earth's history as we study Genesis?

10. How does the Genesis account of our earth's history give us satisfactory explanations for:

    • Fossil fuels such as petroleum and coal.
    • Sedimentary rock formations which are rich in fossils of once living creatures.
    • The abundant evident of glaciers that once covered our earth to a depth of several miles in thickness.
    • Continents that have apparently been pulled part from each other.

11. How does the very existence of fossils refute the uniformitarian view of geology?

12. How many years from the creation of Adam transpired to the time of the flood of Noah?

13. According to verse 16, who shut the door of the Ark when the rain started? Do you see a parallel of this and what God will do in the future according to 2Thessalonians 2:11?

14. How would you explain from Genesis such geologic phenomena as the following:

    Igneous Rock formations
    Sedimentary rock formations as
    • Sandstone

- Shale
- Limestone
- Conglomerates
- Metamorphic Rocks
- Petroleum Deposits
- Coal deposits
- Glacial Geology

15. What promise did God make to Himself in verse 21 of Chapter 8?

16. Are there four seasons in verse 22? What are they? Did these seasons exist before the flood? Might Ecclesiastes 3:15 also relate to this?

# Genesis Annotations Chapter 9

## The Covenant with Noah

### Genesis 9:1-7 (KJV)

[1] And God blessed[a] Noah and his sons, and said unto them, Be fruitful, and multiply, and replenish the earth. [2] And the fear of you and the dread of you shall be upon every beast of the earth, and upon every fowl of the air, upon all that moveth *upon* the earth, and upon all the fishes of the sea; into your hand are they delivered. [3] Every moving thing that liveth shall be meat[b] for you; even as the green herb have I given you all things. [4] But flesh with the life thereof, *which is* the blood[c] thereof, shall ye not eat. [5] And surely your blood of your lives will I require; at the hand of every beast will I require it, and at the hand of man; at the hand of every man's brother will I require the life of man. [6] Whoso sheddeth man's blood, by man[d] shall his blood be shed: for in the image of God made he man. [7] And you, be ye fruitful[e], and multiply; bring forth abundantly in the earth, and multiply therein.

### [a] God Blessed Noah and His Family

God gives to Noah and his family in 9:1&2 the same commission that He gave to Adam and Eve in Gen 1:22. This will be important to remember when we get into Chapters 10 and 11.

God blessed Noah and his sons and said unto them "Be fruitful and multiply and replenish the earth."

- This takes us back to Genesis 1:23 & 24 when God blessed the fish and the fowl to fill the waters of the sea and the earth. Only the word "replenish" regarding the animal and aquatic life is omitted in Genesis 1.
- This same commission (as in Gen 9:1) was given Adam and Eve in Genesis 1:28 along with the "replenish" command. This tells us that the earth was populated by intelligent life between Genesis 1:1 and 1:2. But Adam and Eve were to reign over all of the rest of God's creation. This takes us to Hebrews 2:8 -15 where we see that God's design and intention always was that man (the human race) will ultimately be the eternal custodian of all of God's creation. However, because of the fall of man the only ones who will do that are those who find the fulfillment of that promise in Christ.
- In Genesis 2:3 we see God "blessing the seventh day because that in it He had rested from all his work..." This was not a rest because of weariness but rest in the sense of sitting back and appreciating the master piece of His creation. We can relate to this by reflecting on how we appreciate the conclusion of a job well done or the finishing of a creative piece of work.

  Reflecting back on creation week, we might ask "Why did God take seven days to do something that He could have done in one?" There is surely a reason for it. In Hebrews 4:9 we find a clue to the answer to this question. There we see that there remains "a rest for the people of God." This is a reference to the millennium. If we picture the seven days of creation as a type of God's plan for the ages with one day being as a thousand years as with what Peter says in 2Peter 3:8, then the seventh millennium (coming up soon) would be "the millennium" that Revelation 20:2 talks about. God rested after the six days of creation in Genesis 1 but then had to go to work in the long task of redemption because of the fall of man.

- In Genesis 8:17 we see the parallel to Genesis 1:27 with regard to the animals breeding abundantly in the earth.
- In Genesis 10:32 we see that this blessing in Genesis 9:1 is fulfilled as "by these [the three sons of Noah] were the nations divided in the earth after the flood."

## [b] A Dietary Change

A change in dietary laws is presented in 9:2-3. In Genesis 2:17 we see that the original diet was a vegetarian diet. Now that diet is changed to include the flesh of animals. "Every moving thing that liveth shall be meat for you; even as the green herb have I given you all things." However, catching this food source will be now complicated by the fact that they would now be afraid of man (Genesis 9:2). In Leviticus Chapter 11 we see God again changing the dietary restrictions by separating animals into clean and unclean as to being suitable for food for the children of Israel under the Law of Moses. Then, in 1Timothy 4:3, we see another change – that of God allowing all food to be received with thanksgiving by believers today in the dispensation of grace.

## [c] Blood was not to be consumed

We find in Genesis 9:4-5 that blood was not to be consumed.

- There was something special about the blood. The blood that was shed to cover sin was the essential ingredient of the sacrifice that God required. The blood shed by an innocent animal was the issue between Cain's and Abel's offering in Genesis 4. Under the Law of Moses, Israel was to eat neither fat nor blood because both of these were involved in the sacrifices of the altar (Leviticus 3:17; 7:26; 17:10-14; 19; 36 etc al.).
- In Acts 15:20-29, we see that the circumcision believers in Jerusalem requested that Gentile believers in the Dispensation of Grace not eat anything strangled (because the blood was still in it) or to consume blood in any form. This was apparently in regard to the sensitivities of the circumcision believers during the Acts period because they (the circumcision believers) were still under the law during the Acts period.
- However, the normal course for the Dispensation of Grace is that "…every creature of God is good, and nothing is to be refused, if it is received with thanksgiving." The consumption of blood is no longer an issue because the animal sacrifices have no value anymore now that the blood of Christ has ended the need for the blood sacrifices in that now we have "…Redemption through his blood the forgiveness of sins…" (Ephesians 1:7; Colossians 1:14; Hebrews 9:12).

## [d] Human Government is Instituted

The institution of human government is presented in this chapter. Genesis 9:5 speaks of society taking action to avenge the taking of a human life. This requires the organization of a government with a legal system. In Genesis 4:9 and forward we see that when Cain killed Abel that Cain was not killed to avenge Abel's death but his life was actually protected (Genesis 4:15). However, all that was changed after the flood. Now whether man or beast kills a man, that man or beast was to be killed. In Exodus 21:12 we see this capital punishment apply to beasts as well as man. In Numbers 35:31-44 we find the interesting account of the avenger of blood. In Romans 13:1-7 we see that capital punishment is a part of this present Dispensation of Grace. "The powers that be are ordained of God…" (Rom. 13:1). God determined that Government has the power to administer capital punishment. This provides for protection for society. The right to punish offenders who would take or threaten human life provides for protection for social order:

--Internal protection (capital punishment)
--External protection (warfare to defend the sovereignty of nations)
--Common laws to protect property rights of people.

In Genesis 10:5 we see that it was God's design that there be separate nations -- each being sovereign. The sovereignty of nations is essential to the safeguarding of the interests of God among the nations of the earth. With nations being sovereign and independent of each other, the laws of one (whether good or bad) does not affect the neighboring nations. Therefore, when one nation goes bad with respect to personal freedom and safety and becomes oppressive, people can seek refuge by fleeing to the neighboring nation where such freedom still exists. We will study this issue more when we come to Chapters 10 and 11 of Genesis.

Nowhere in God's dealings with men was the possession of weapons for defense forbidden. The problem in societies today is not the possession of guns and weapons but the lack of (or the failure of) a system of justice that enforces capital punishment as it was ordained by God.

## [e] Man is Re-Commissioned to Replenish the Earth

Genesis 9:7 is the command to the entire human race to "be fruitful and multiply and bring forth abundantly in the earth and multiple therein" This command has never been rescinded. From the 1960's through 1990's, the philosophies of man were telling people to take action to limit the human population. Now in the 2020s there are men who are going so far as to suggest that the world should be depopulated. However, much of the problem that societies face today is due to the fact that man is not reproducing himself. A society has to have a birth rate of at least 2.1 per couple to maintain itself population wise. Most western cultures today have an average birth rate much lower than that. When a culture or society has a birth rate that drops below 1.7 per couple, it is essentially mathematically impossible for that country to recover from it. In our country for example, we have a birth rate of about 1.7. We are growing at an average rate of about 2.2 per couple per year but only because of immigration (largely Hispanic). Our country will find itself in big trouble economically and demographically because of the declining birth rate. The horrible sin of abortion will come back to haunt our society in more ways than one.

## Genesis 9:8-11 -- The Noahic Covenant:

> [8] And God spake unto Noah, and to his sons with him, saying, [9] And I, behold, I establish my covenant[f] with you, and with your seed after you; [10] And with every living creature that *is* with you, of the fowl, of the cattle, and of every beast of the earth with you; from all that go out of the ark, to every beast of the earth. [11] And I will establish my covenant with you; neither shall all flesh be cut off any more by the waters of a flood; neither shall there any more be a flood to destroy the earth.

## [f] The Details of the Noahic Covenant

Here (in verse 9) we see the covenant that God made with Noah. This is one of eight covenants that God made with man. See Appendix 25 on the Covenants for more detail.
This covenant (Genesis 9:9) is referred to hereafter as the Noahic Covenant. There are a number of aspects to it:

- God establish it with Noah and with his seed. We are a part of the seed of Noah (vs. 9).
- This covenant extends even to every living thing that went out of the ark (vs. 10).
- This covenant involves a promise from God that He will never again wipe out all flesh by the waters of a flood (vs. 11). Nor will He ever destroy all of the earth with a flood. However, we do find in 2Peter 3:10 that one day the elements of the earth will "…melt with fervent heat." This we understand to take place in the future as a part of God's plan to make a new heaven and a new earth (Revelation 21:1). This will be necessary for a number of reasons. For one thing, God will have to repeal the Second Law of Thermodynamic (i.e. the decay principle) since the new heaven and the new earth will last throughout all of eternity. See note [a] on Genesis Chapter 1 for a refresher on this Second Law of Thermodynamics. Another reason for making all things new is that God will one day remove all of the stain of sin from His creation out there in the dispensation of the fullness of times (Eph. 1:10). God has a vested interest in His creation and will therefore "preserve man and beast" (Psalm 35:6).

## Genesis 9:12-17 The rainbow token of the covenant

[12] And God said, This *is* the token[g] of the covenant which I make between me and you and every living creature that *is* with you, for perpetual generations: [13] I do set my bow in the cloud, and it shall be for a token of a covenant between me and the earth. [14] And it shall come to pass, when I bring a cloud over the earth, that the bow shall be seen in the cloud: [15] And I will remember my covenant, which *is* between me and you and every living creature of all flesh; and the waters shall no more become a flood to destroy all flesh. [16] And the bow shall be in the cloud; and I will look upon it, that I may remember the everlasting covenant between God and every living creature of all flesh that *is* upon the earth. [17] And God said unto Noah, This *is* the token of the covenant, which I have established between me and all flesh that *is* upon the earth.

## [g] The Sign of the Covenant..

Each of the covenants that God has made has a token (or a sign). This covenant has the rainbow as its token. Because of the different hydrologic cycle that existed before the flood of Noah, the rainbow was not produced in that it did not rain in the pre-flood world (Genesis 2:5). The rainbow will be "for perpetual generations…" It will always be a part of the normal functioning of the earth from that time forward (verse 12). This Noahic covenant is an "everlasting covenant" (vs. 16), The rainbow is an interesting representation as a token of the Noahic Covenant,. Ezekiel describes the throne of God in Ezekiel 1:28 -- "As the appearance of the bow that is in the cloud in the day of rain, so was the appearance of the brightness round about." This was "… the appearance of the brightness of the glory of the LORD…" When John saw the Lord in Revelation 4:3, he describes Him as follows: "…he that sat was to look upon like a jasper and a sardine stone. And there was a rainbow round about the throne insight like unto an emerald." The mighty angel of Revelation 10 was "…clothed with a cloud: and a rainbow was upon his head, and his face was as it were the sun" (Revelation 10:1). We note that the rainbow is a remembrance for God for He said "…when I bring a cloud over the earth, that the bow shall be seen in the cloud: And I will remember my covenant… and I will look upon it, that I may remember the everlasting covenant between God and every living creature…" (Genesis 9:14-16)

A number of the covenants God made had such a token or sign.
  --The sign of the Noahic Covenant was the rainbow
  –The sign of the Abrahamic Covenant was circumcision (Romans 4:11)
  --The sign of the Mosaic Covenant was the Sabbath (Exodus 31:13)
  --The sign of the Davidic Covenant was water baptism (Lev.26:40-42
      David's seed was to be the king over a kingdom but this kingdom was to be a kingdom of priests and a royal priesthood. Water baptism, being the inductory rite into the priesthood, became the sign of this covenant. Israel will be a kingdom of priests but not until after they come to the baptism of repentance (Lev. 26:40-42; Mark 1:4; Luke 3:3; Acts 13:24; 19:4)

## Genesis 9:18-23 - The Sons of Noah

[18] And the sons of Noah, that went forth of the ark, were Shem, and Ham, and Japheth: and Ham *is* the father of Canaan. [19] These *are* the three sons of Noah: and of them was the whole earth overspread. [20] And Noah began *to be* an husbandman, and he planted a vineyard: [21] And he drank of the wine, and was drunken[h]; and he was uncovered within his tent. [22] And Ham, the father of Canaan, saw the nakedness of his father, and told his two brethren without. [23] And Shem and Japheth took a garment, and laid *it* upon both their shoulders, and went backward, and covered the nakedness of their father; and their faces *were* backward, and they saw not their father's nakedness.

## [h] Noah the Husbandman

Verses 18 through 23 are discouraging to read after reading about how God brought Noah and his family safely through the flood. We find Noah began to be a husbandman and planted a vineyard. He apparently had not been a husbandman (one who tends to plants and animals – i.e. agriculture) before. For the past year he was a sailor and for 120 years before that he was a carpenter and a shipbuilder. But here we see Noah growing vines and making wine. There was nothing wrong with growing grapes or with making wine – which was a means of preserving the harvest. The problem came from being drunk with wine (Eph. 5:18). While being passed out and drunk in his tent, he awakes and realizes that he had been greatly disrespected at best. It did not take long for him to figure out who did it for he knew his sons.

But it is interesting that Noah curses Canaan the son of Ham. Why (we ask) would he not curse Ham? One obvious reason is that one cannot curse some one who God has blessed and we note that "God blessed Noah and his sons…" (9:1) Here we see a man of whom God said "…you have I seen righteous before me in this generation…" drunken to the point of passing out. How true it is that "wine is a mocker, strong drink is raging, and whosoever is deceived thereby is not wise." (Proverbs 20: 1; 23: 31 & 32) Worse yet, we see the son of such a man violate the person of his own father. Again this brings to mind the truth of John 8:34 "Whosoever committed sin is a servant of sin."

The apostle Paul says it in Romans 6:16 "Know ye not that to whom ye yield yourselves servants to obey, his servant ye are whom ye obey; whether to sin unto death and of obedience unto righteousness?"(Romans 6:16) Ham served sin -- the sin nature that resides within every man. Sin always pays its wages of death (Romans 6:23). Noah knew Canaan and Canaan's inclination and he knew from whom he had gotten it. This bears out the Scripture that "Be not deceived; God is not mocked: for whatsoever a man soweth, that shall he also reap. For he that soweth to his flesh shall of the flesh reap corruption; but he that soweth to the Spirit shall of the Spirit reap life everlasting." (Gal 6:7-8) It might well be that Ham's son Canaan was already grown and was with him at the time.

## [i] Parents are to be Honored and Respected

Any disrespect of parent is a serious thing with God. We see that in passages such as Leviticus 18:6, 7, 24, 29, 30; 20:9, 22-24; and Deuteronomy 9:4; 12:31; et. al. We note in verse 18 that the only descendant of any of the three sons of Noah named is Canaan the son of Ham. God singles out Canaan here for a reason. We will see noted in the next chapter that five of the nations that were to be destroyed in the land of Canaan were descendants of Canaan (Gen.10:16-17). Note [j] below discusses why they were to be destroyed.

## Genesis 9:24-29 (KJV)

24 And Noah awoke from his wine, and knew what his younger son had done unto him. 25 And he said, Cursed[i] *be* Canaan; a servant of servants shall he be unto his brethren. 26 And he said, Blessed *be* the LORD God of Shem; and Canaan shall be his servant. 27 God shall enlarge Japheth, and he shall dwell in the tents of Shem; and Canaan shall be his servant. 28 And Noah lived after the flood three hundred and fifty years. 29 And all the days of Noah were nine hundred and fifty years: and he died.

## [j] Canaan Cursed by Noah

In verse 25 we see Noah put a curse on Canaan his grandson. Because he could not curse Ham whom God had blessed, he cursed the off spring in which the same heart attitude resided. The descendants of Canaan are cursed to serve the descendants of Shem and Japheth. This happened in 1Kings 9:20 & 21 and 2Chronicles 8: 7 & 8 under Solomon. It should be noted also that this cursing of Canaan was actually prophetic for it did not come about until about 800 years later. God who knows the future foresaw that the descendants of

Canaan (the Canaanites) would be the people that Satan would use to populate the land of Canaan with the race of giants in order to keep the children of Israel from inheriting the land promised to Abraham (Gen 12:7). However, God is patient and does not exclude anyone from the possibility of repentance. God waited with the descendants of Canaan until their iniquity was full (Gen 15:16). We remember the study that we made of the sons of God taking wives of the daughters of men to produce giants and noted that "There were giants in the earth in those days [i.e. before the flood]; and also after that [i.e. after the flood], when the sons of God came in unto the daughters of men, and they bare *children* to them, the same *became* mighty men which *were* of old, men of renown." (Gen 6:4) Under the conquest of the land, the Canaanites were either put to the sword or were put to tribute (Joshua 9:23; Judges 1:28-35).

Study Guide Questions for Chapter 9

1. In verses 2 and 3 we see the diet of man changed from what it was before the flood. Describe that change.

2. What according to verses 4 and 5 was not to be eaten?

3. Verse 5 introduces a new institution to human society. What was that institution? Does this still affect us today?

4. What command does God give man in verse 7? Has this command ever been rescinded?

5. What Covenant do we find in verses 7 through 11?

6. What is the sign or the token of the Noahic Covenant?

7. In verse 25 we see Noah put a curse on Canaan. What was that curse and why was it placed?

# Genesis Annotations Chapter 10

## The Generations of Noah

## Genesis 10:1 (KJV)

[1] Now these *are* the generations[a] of the sons of Noah, Shem, Ham, and Japheth: and unto them were sons born after the flood.

## [a] The Generations of the Sons of Noah

All of the families of the nations of the earth today are descendents of the children of Noah. Though it is not possible to accurately track the dispersions of the children of Noah in their travels, and their distributions over the earth, historians, anthropologists, and archeologists have none the less done a good job of tracking the migrations of the various peoples of the earth. The attached structural analysis of chapter 10 of Genesis is laid out as a family tree of these three sons of Noah. Next to each of the patriarchs are the region and/or nations and peoples of the world today that represent their descendants. I will not say that this is entirely accurate in all of its detail nor is it all that important to accurately know all of that information. However, it is none the less interesting to know what one's roots are in the family of nations as to origins in the Book of Genesis. Much of the information here was obtained from the web page www.biblebelievers.com. A more detailed look is available from the book *General Biblical Introduction – From God to Us* by H. S. Miller who majored in the study of ethnology.

## Genesis 10:2-5 (KJV) - The Sons of Japheth

[2] The sons of Japheth; Gomer, and Magog, and Madai, and Javan, and Tubal, and Meshech, and Tiras. [3] And the sons of Gomer; Ashkenaz, and Riphath, and Togarmah. [4] And the sons of Javan; Elishah, and Tarshish, Kittim, and Dodanim. [5] By these were the isles of the Gentiles divided in their lands; every one after his tongue, after their families, in their nations.

## Genesis 10:6-20 (KJV) - The Sons of Ham

[6] And the sons of Ham; Cush, and Mizraim, and Phut, and Canaan. [7] And the sons of Cush; Seba, and Havilah, and Sabtah, and Raamah, and Sabtecha: and the sons of Raamah; Sheba, and Dedan. [8] And Cush begat Nimrod[b]: he began to be a mighty one in the earth. [9] He was a mighty hunter before the LORD: wherefore it is said, Even as Nimrod the mighty hunter before the LORD. [10] And the beginning of his kingdom was Babel, and Erech, and Accad, and Calneh, in the land of Shinar. [11] Out of that land went forth Asshur, and builded Nineveh, and the city Rehoboth, and Calah, [12] And Resen between Nineveh and Calah: the same *is* a great city. [13] And Mizraim begat Ludim, and Anamim, and Lehabim, and Naphtuhim, [14] And Pathrusim, and Casluhim, (out of whom came Philistim,) and Caphtorim. [15] And Canaan begat Sidon his firstborn, and Heth, [16] And the Jebusite, and the Amorite, and the Girgasite, [17] And the Hivite, and the Arkite, and the Sinite, [18] And the Arvadite, and the Zemarite, and the Hamathite: and afterward were the families of the Canaanites spread abroad. [19] And the border of the Canaanites was from Sidon, as thou comest to Gerar, unto Gaza; as thou goest, unto Sodom, and Gomorrah, and Admah, and Zeboim, even unto Lasha. [20] These *are* the sons of Ham, after their families, after their tongues, in their countries, *and* in their nations.

## [b] Chapter 10 is a Genealogy of Noah's Family

Except for two portions of Chapter 10 (the mention of Nimrod in verses 8 – 11 and the mention of Peleg in verse 25); all of Chapter 10 is a genealogy of Noah's family. Though these two digressions appear quite insignificant and mundane in the text, there is great significance to these simple statements as far as our world today is concerned. Most of Chapter 10 covers things that happened chronologically after the events of Chapter 11. However, these two passages serve as an introduction to Chapter 11. The passage in verses 8 thru 12 regarding Nimrod will be discussed at some length in Chapter 11. However, the passage in verse 25 regarding Peleg is worth some additional consideration.

**Genesis 10:21-24 (KJV) The Sons of Shem**

21 Unto Shem also, the father of all the children of Eber, the brother of Japheth the elder, even to him were *children* born. 22 The children of Shem; Elam, and Asshur, and Arphaxad, and Lud, and Aram. 23 And the children of Aram; Uz, and Hul, and Gether, and Mash. 24 And Arphaxad begat Salah; and Salah begat Eber.

**Genesis 10:25-30 (KJV) The Sons of Eber**

25 And unto Eber were born two sons: the name of one *was* Peleg[c]; for in his days was the earth[d] divided; and his brother's name *was* Joktan. 26 And Joktan begat Almodad, and Sheleph, and Hazarmaveth, and Jerah, 27 And Hadoram, and Uzal, and Diklah, 28 And Obal, and Abimael, and Sheba, 29 And Ophir, and Havilah, and Jobab: all these *were* the sons of Joktan. 30 And their dwelling was from Mesha, as thou goest unto Sephar a mount of the east

**[c] The Earth Divided Into Continents**

The name Peleg in verse 25 means "division" and he was so named because "…in his days was the earth divided." It should be noted that Peleg was in the godly line of Shem. The lineage of Peleg is as follows:
Shem-> Arphaxad->Salah->Eber->Peleg &
->Jaktan

This division that verse 25 speaks about is not the division of the population of the earth into families, nations, and tongues that Chapter 11 will address. Rather, this division of the earth is speaking about the division of the physical earth as in the separation of the continents of the earth. If we study a globe, we see that the continents can be easily moved together to fit back to what would have been one land mass. Geologists call it Pangaea. This would have been the land mass onto which Noah and his family emerged when they left the ark. As we study the physical earth, we see features that can only be explained by some cataclysmic event that physically pulled the various continents apart from each other. As we study Genesis 11, we see why God would precipitate such an event. In Genesis, 11, we will see that man will be organizing himself in a one world government system that would exclude God from the planet (Gen. 11:3 & 4). God countered by the confusion of tongues to cause the various families of the human race to have to separate from each other. Then, some time later, He precipitated an event that separated the continents to further separate nation from nation. Peleg was born in 1757 A.M. or about 2218 B.C. Peleg was born about the time of the Tower of Babel. It is likely that he was named Peleg (literally "division") as a prophecy given by God to Eber (Peleg's father). The tower of Babel (built by Nimrod) was probably built shortly before the birth of Peleg. The family line to Nimrod is as follows:
Ham->Cush->Nimrod.
Nimrod was the first post flood world emperor. In fact, there has not been another world leader that ever had as broad a scope of dominion as did Nimrod. He had complete control over the entire world population. Many leaders since then have desired and actively sought after what Nimrod had but none have ever accomplished it. The next leader that will come close to accomplishing that will be the antichrist. Though the antichrist will not directly control all of the nations of the earth, he does influence all of the nations in that

he is able to martial all of the armies of the earth to fight against the Lord when He returns to set up His kingdom (Rev. 19:19-20).

Lloyd Nolen Jones in his book "*The Chronology of the Old Testament*" (2019) suggests that Nimrod's kingdom began in c. 1822 AM (*Chronology of the Old Testament* p. 42). This date being 166 years after the flood would result in a human population of over 30,000 people (his estimate). This estimate assumes 30 years per generation and twelve children per generation. The life span of Peleg is from 1757 AM to 1996 AM for a total of 239 years. The confusion of tongues that separated families from each other and stopped the rebellion likely would have occurred early in that timeframe (perhaps c. 1850AM) with the division of the land mass of Pangaea discussed here in Chapter 10 being close to the end of the 239 years (c. 1996AM). See the time line presentation in Appendix 1 for a perspective of the relative events with respect to time and dates.

## [d] Genesis and Geology

We had looked at how the original creation as described in Chapter 1 accounts for the basic igneous bedrock geology that we find in our earth today. Appendix 5 covers this. We saw also that the sedimentary geology (i.e. the sandstones, shales, and dolomites of the earth's surface) were deposited as a result of the flood of Noah's day. Yet there are many geologic features that we observe in the earth for which we need (or at least desire) an explanation. Those include the soils and geologic features that we call glacial features such as drumlins, eskers, kettle holes, vast gravel deposits, outwash sand deposits, and much of the soils material that we find in Wisconsin, Minnesota, Michigan, New York, etc. We can add to that the wooly mammoths that were frozen so fast that they did not even have time to swallow their last meal of warm climate vegetation. Also, we might consider the mountain ranges that have been pushed up to greater heights than could be possible by forces that exist on this earth alone. Associated with the mountain ranges, we might consider the related phenomenon of the mid ocean rift valleys and the great rift valley of east Africa. As we consider these phenomenons and consider that Genesis and the Scriptures must give us the answers to the origin of the earth, we look to the Scriptures for explanations for these geologic features as well.

There is a passage in the Book of Job that makes a very interesting reference to ice that might be of some real help to us in understanding Genesis 10:25.

> [29] Out of whose womb came the ice? and the hoary frost of heaven, who hath gendered it? [30] The waters are hid as *with* a stone, and the face of the deep is frozen. (Job 38:29-30)

We find this passage in a context that speaks of God's providential work in regard to the normal course of nature. This passage explains both terrestrial and astrological phenomena that man cannot explain without understanding that God was responsible for them. The 38th chapter of Job is rich with scientific information that man could learn from if he would but let these passages speak to him. We ask ourselves: What is this ice which had come from the womb of God? And what is this hoary frost from heaven? And in what sense is the deep frozen?" If we take this passage literally that there was ice that came from heaven, then we are on the verge of understanding what it is that produced the glaciers that left their impressive mark on our earth.

It is interesting to study what theories man (specifically men who disbelieve the Bible) comes up with to explain these phenomenon by natural and uniformitarian processes. Uniformitarianism is the belief held by most scientists that the processes that have formed the geologic features of this earth through the ages are the same uniform processes as are at work through natural processes in the earth today. Genesis however has a different explanation for the earth's geology. Genesis presents a series of cataclysmic events in explaining what we observe in the earth's soils and geologic features.

It is estimated by scientists studying the amount of material transported by the glaciers that these masses of ice had to be about 10,000 feet thick. In some of my work in the field of engineering over the years, I had encountered often the striations in the bedrock surface that were left by the glaciers as they passed over the bedrock. In eastern Wisconsin, we had encountered black ash forests that had been overflowed by glaciers and buried by 20 to 50 feet of glacial deposits. Being buried deep enough to preserve the vegetation but not under so much pressure that the material was turned to coal as was done in the case of the vegetation buried by the sediments from the flood. That vegetation would be in the form of the original trees that were buried. When that buried vegetation was brought up to the surface, it would very quickly decompose to an ash almost overnight. This is the result of it being exposed to the oxygen rich atmosphere once it is pulled up to the surface. Those forests buried by the glacial deposits were on top of sedimentary material that would have been deposited by the flood of Noah's day. This indicates that there was a fairly long time delay between the flood and the glaciers.

The theories that are put forth by unbelieving scientists as explanations for the ice ages are for the most part devised and presented specifically to attempt to disprove the Genesis account. There is indeed in the world today what the apostle calls "… the oppositions of science falsely so called" which he warns the young man Timothy to watch out for and to be on guard against (1Timothy 6:20).  Others are presented by Bible believers to try to associate the ice ages with the Genesis account of the flood. However, the observable information makes it evident that the two events (the flood deposition and the deposition of glacial material were separated by some time. That is to say that the actions that produced the flood and the forces that produced glaciers and the physical division of the earth into continents were not the same forces.

There is one explanation that does do justice to the facts and presents an explanation that allows the account to be taken literally with out making any forced application of natural phenomena that is derived from the natural weather patterns that we see in the world today. That explanation was presented by Donald W Patton in his book *The Biblical Flood and the Ice Epoch* published by the Pacific Meridian Publishing Company. In that book, Mr. Patton takes the view that the ice that came from God and the "hoary frost from heaven …" (Job 38:29) was indeed from God and from heaven. If we were to take this ice as being an asteroid of ice that came into the proximity of the earth from outer space, it would have arrived to the earth's environs at a temperature of near absolute zero (minus 470 degrees Fahrenheit -- zero degree Rankin or minus 273 degrees centigrade – zero Kelvin).

Mr. Patton was well educated in astronomy and was in fact an instructor and lecturer in astrophysics. He presents a model of what would happen if an asteroid of ice would pass through the plane of the earth's orbit at an angle of 45 degrees or more to the plane of the earth's orbit. There is a law in astrophysics called Roche's law. That law states that as two bodies approach each other in space, at a distance of about 2.44 times the diameter of the larger body, the gravitational pull of the larger body would disintegrate the smaller body. At the same time, the gravitational pull of the smaller body would cause significant changes in the larger body. If earth was affected by an asteroid of ice in this manner, we would have an explanation for all of the otherwise unexplained phenomenon listed above. The earth, in this scenario, would be the larger body and the asteroid of ice would be the smaller.

Let's consider the mechanics of such a possibility. To follow this discussion, you will have to direct your attention to Figures 3 through 8 appended to the end of this chapter. These are sketches contained in the book *"The Biblical Flood and the Ice Epoch"* by Donald W. Patton. The book is out of print but copies are still available. I encourage anyone interested in pursuing this study further to get a copy of this excellent reference on the subject. Looking at Figure 6, you see an illustration of the earth's orbit, the earth's electro-magnetic belts (i.e. the Van Allan Belts) and the earth's Roche's limit. The Van Allen Belts of the earth are also illustrated on Figure 5. These belts are electro-magnetic belts that exist around our earth. Figure 4 illustrates how the earth is oriented relative to its path and the orientation of the magnetic belts. Now let's consider what would happen if an asteroid of ice disintegrated into ice particles under the influence of the gravitational pull of the earth and then be affected by the Van Allen belts.

As the ice particles bounced around against each other, they would be giving up electrons and others receiving electron much as ice particles do in our atmosphere to produce the lightening and thunder in a typical thunder storm. Once the ice is disintegrated to smaller particles, they would not be subject to quite as much of a pull to the earth as with the large ice mass. However, being now charged, they would be drawn to the magnetic poles of the earth around the Van Allen belts. The positively charged particles would be drawn to the South Pole while the negatively charged particles to the North Pole. The result would be that the ice mass would fall to the earth with its center at the magnetic poles. Sheet 1 of 6 (Figure 3) shows the location of the ice masses in the northern hemisphere relative to the geographic North Pole and the magnetic north pole.

Let's go further with the consideration of this model in regard to the mountain building processes that produced the mountain ranges and the rift valleys of the earth's crust. Considering this, I direct your attention to Figure 8. There are four forces involved in the process:
1. The tidal effect of the asteroid of ice on the molten core of the earth to produce tides in the core much like the moon does on our oceans.
2. The earth's gravity acts on the asteroid of ice to disintegrate it.
3. The earth's centrifugal force together with its angular momentum acts to tear the crust of the earth open at north to south arcs.
4. The asteroid's gravitational pull causes the earth to bulge at its equatorial dimension.

The process is understood as one visualizes the earth becoming oblong due to the tidal effect on the earth's molten core. The earth's circumference increases at the equator and thus the crust cracks at the seams of the various plates that we today call the tectonic plates of the earth. As the plates on the earth's surface tend to be slowed down by the gravitational pull of the visitor and the momentum of the earth's core continue to spin normally, a great drag force is exerted on the plates which then are pulled apart from each other. The trailing side of one plate would then tend to ride over or under the leading edge of the plate that follows it. Thus the force to raise mountains to higher elevations than could be possible from forces that exist on the earth today is understood. So too, the volcanoes that comprise the Pacific "Ring of Fire" can be easily visualized as the molten core of the earth is opened to the surface at the seams of the plates. We can now visualize how the physical earth was divided into continents in the days of Peleg. God not only separated the families of the earth from each other by their tongues, he also pulled the very continents on which they lived apart.

For those interested in a more descriptive explanation of the asteroid of ice and the effect it would have on the earth, there is a video "*Cataclysm from Space 2800 BC*" available on You Tube. In that video, Donald W. Patton presents in graphic form the theory that he presents in his book "*The Biblical Flood and the Ice Epoch.*" It should be noted however, that Donald Patton sees the Ice Epoch that divided the continents from each other being coincident with the flood of Noah's day while this author sees it as occurring about a century later in the days of Peleg.

The result of all of this is that the earth is divided into continents as we see it today. Also we can also understand how the wooly mammoths were frozen so quickly and so thoroughly that they are occasionally discovered today in Siberia as the ice melts to reveal their frozen flesh. We can also now understand how the glacial ice could advance so fast on this earth. For the believer, we can be satisfied that we do not need to fear the attacks that unbelief launches on the Bible. We can let Scripture say what it says. We can then focus on getting to the real issue of the Scripture – that of the presentation of the redemption that is available to all men in the work that the creator accomplished on the cross when He personally settled the account for the sin's debt on behalf of all men.

## Genesis 10:31-32 - The Sons of Shem

[31] These *are* the sons of Shem[e], after their families, after their tongues, in their lands, after their nations. [32] These *are* the families of the sons of Noah, after their generations, in their nations: and by these were the nations divided in the earth after the flood.

## [e] The Sons of Shem

Verse 31 picks up with the generation of Shem. The Bible leaves off with the Gentiles here at Chapter 10 and focuses on the descendants of Shem and the seed line through Abraham to the Messiah of Israel. The families of the earth are divided "...after their families, after their tongues, in their lands, after their nations." God confounds the languages of the human race in Genesis 11 and thereby causes the various families to separate form each other. We will see why in Chapter 11. Once God had separated the nations from each other, He further separates them by dividing the earth and their lands from each other by the mountain building processes that formed the earth as we have it today. .

## [f] Farewell to the Gentiles

From the close of Chapter 11, God does not pick up His dealings with the Gentiles until He saves Saul of Tarsus in the Book of Acts and sends him to the Gentiles with the mystery program (Eph. 3:1-3). It is through Paul that the Lord Jesus reveals to the world the information regarding the dispensation of grace in which we live today and relate to God as Gentiles who are now in a position to receive grace to be saved.

### Acts 17:22-31

[22] Then Paul stood in the midst of Mars' hill, and said, *Ye* men of Athens, I perceive that in all things ye are too superstitious. [23] For as I passed by, and beheld your devotions, I found an altar with this inscription, TO THE UNKNOWN GOD. Whom therefore ye ignorantly worship, him declare I unto you. [24] God that made the world and all things therein, seeing that he is Lord of heaven and earth, dwelleth not in temples made with hands; [25] Neither is worshipped with men's hands, as though he needed any thing, seeing he giveth to all life, and breath, and all things; [26] And hath made of one blood all nations of men for to dwell on all the face of the earth, and hath determined the times before appointed, and the bounds of their habitation; [27] That they should seek the Lord, if haply they might feel after him, and find him, though he be not far from every one of us: [28] For in him we live, and move, and have our being; as certain also of your own poets have said, For we are also his offspring. [29] Forasmuch then as we are the offspring of God, we ought not to think that the Godhead is like unto gold, or silver, or stone, graven by art and man's device. [30] And the times of this ignorance God winked at; but now commandeth all men every where to repent: [31] Because he hath appointed a day, in the which he will judge the world in righteousness by *that* man whom he hath ordained; *whereof* he hath given assurance unto all *men*, in that he hath raised him from the dead." (Acts 17:22-31 KJV)

See Appendix 25 "The Covenants of God and the Mystery" for a brief introduction to the mystery. For a more complete presentation and discussion of the mystery, see the book *You and Your Creator* (Tiry, 2021).

Structural Layout: Genesis 10

The generations of Noah

[1] Now these *are* the generations of the sons of Noah, Shem, ... The Semitic People of the Middle
East
Ham, ...      The Colored Peoples
and Japheth:  The Indo-European People

and unto them were sons born after the flood.


[2] The sons of Japheth; Gomer, ...     The Celtic People, French, Germans, Scandinavian, Brittons,
Welsh, Dutch, Romans
and Magog, ...   Georgian, Russian, Scythians, Belarusian, Ukrainian, Turks,
and Madai, ...   Medes, Kurds, Persians, East Indians
and Javan, ...   Ionians, Greeks, Cypriots, Spanish, Italians, Portuguese
and Tubal, ...   Bulgarians, Hungarians, Albanians, Romanians
and Meshech,   Northern Turkey, Poles, Finns, Czechs, Yugoslavia
and Tiras.      Tracians, Etruscans, Italy, Western Turkey


[3] And the sons of Gomer; Ashkenaz, ...   Armenia, Bithynia, Western Germany, Scandinavia, Denmark
and Riphath, ...   Czechoslovakia, Romania, Turkey
and Togarmah.   Armenia, Turkey, Germany


[4] And the sons of Javan; Elishah, ...   Greece, Sicily
and Tarshish,   Spain
Kittim, ...   Cyprus
and Dodanim.  Rhodes and the Dardanelle


[5] By these were the isles of the Gentiles divided in their lands; ...
every one after his tongue, ...
after their families, ...
in their nations.

The sons of Ham
[6] And the sons of Ham; Cush, ...   Arabia
and Mizraim, ...   Egypt, Lydia
and Phut, ...
and Canaan.   Lebanon, Phoenicia, N Africa


[7] And the sons of Cush; Sheba, ...   Arabia, Yemen
and Havilah,   Northern Arabia
and Sabtah, ...   SE Arabia
and Raamah, ...   SE Arabia
and Sabtecha: ...   SE Arabia
and the sons of Raamah; ...
Sheba, ...   North of Yemen
and Dedan.   United Arab Emirates


119

[8] And Cush begat Nimrod: …

> he began to be a mighty one in the earth.
> [9] He was a mighty hunter before the LORD: …
> wherefore it is said, Even as Nimrod the mighty hunter before the LORD.
> [10] And the beginning of his kingdom was Babel, … in the land of Shinar
> > and Erech, …
> > and Accad, …
> > and Calneh,.
> > [11] Out of that land went forth Asshur, …          Assyria
> > and builded Nineveh, …
> > and the city Rehoboth, …
> > and Calah,
> > [12] And Resen …
> > > between Nineveh and Calah:…
> > the same *is* a great city.

[13] And Mizraim begat Ludim, …                                    Lydia

and Anamim, …                                    The Nile Region
and Lehabim, …                                    Lydia
and Naphtuhim,                                    North Nile Delta
[14] And Pathrusim, …                                    Upper Egypt
and Casluhim, (out of whom came Philistim,) …                                    N Egypt
and Caphtorim.                                    The Philistines

[15] And Canaan begat Sidon his firstborn, …          N. Africa
and Heth,                                    Carthage area, Hittites, Cathay region of China
[16] And the Jebusite, …
and the Amorite, …          East of Canaan (Josh 24:11)
and the Girgasite,          East of the Jordan River (Exodus 15:16)
[17] And the Hivite, …          Mt. Hermon (Gen 34:2; Josh 9:7-17; 11:19)
and the Arkite, …          Lebanon
and the Sinite,          The Sino people of China, Native  American, South Seas
[18] And the Arvadite, …          Phoenician
and the Zemarite, …          Coast of Lebanon
and the Hamathite: …          Upper Syria
and afterward were the families of the Canaanites spread abroad.
[19] And the border of the Canaanites was from Sidon, as thou comest to Gerar, …
unto Gaza; as thou goest, unto Sodom, …
and Gomorrah, …
and Admah, ….
and Zeboim, even unto Lasha.

[20] These *are* the sons of Ham, after their families, …
after their tongues, ….
in their countries, …
*and* in their nations.

120

The sons of Shem

²¹ Unto Shem also, the father of all the children of Eber, ...Father of the Hebrews
the brother of Japheth the elder,...
even to him were *children* born.

²² The children of Shem; Elam,     East of Babylon an north of Persia
and Asshur,...     Assyrians north of Iraq
and Arphaxad,     Father of the Chaldeans, Hebrews, Arabians, Amorites,
    Moabites, Jordanians
and Lud,...     Lydia
and Aram.     Father of the Aramaeans of Syria and Lebanon
²³ And the children of Aram; Uz,...     Land of Job (Job 1:1)
and Hul,...
and Gether,...
and Mash.

²⁴ And Arphaxad begat Salah;...
and Salah begat Eber.
²⁵ And unto Eber were born two sons:...
the name of one *was* Peleg; for in his days was the earth divided;... Name: "division"
and his brother's name *was* Joktan.

²⁶ And Joktan begat Almodad,...     Very little trace
and Sheleph,...     Yemen
and Hazarmaveth,...     SE Arabia
and Jerah,     SE Arabia
²⁷ And Hadoram,...     S Arabia
and Uzal,...     Yemen

and Diklah,     NE Arabia
²⁸ And Obal,...     Yemen
and Abimael,...     Yemen
and Sheba,     Yemen
²⁹ And Ophir,...     SW Arabia
and Havilah,...     East Shore of the Gulf of Aqaba
and Jobab:     Area around Mecca
all these *were* the sons of Joktan.

³⁰ And their dwelling was from Mesha, as thou goest unto Sephar a mount of the east.

³¹ These *are* the sons of Shem, after their families,...
after their tongues,...
in their lands,...
after their nations.

³² These *are* the families of the sons of Noah, after their generations,...
in their nations:...
and by these were the nations divided in the earth after the flood.

Study Guide Questions on Chapter 10

1. In verse 25 we see one of the Patriarchs whose name is division. What division was he named for?

2. How has that division shaped the world today?

3. What features in the geology of the earth today has resulted from that event that produced the division of the earth?

4. Why does the chapter close with a focus on the generations of Shem?

GLACIOGENESIS: THE CAUSE OF THE ICE EPOCH 113

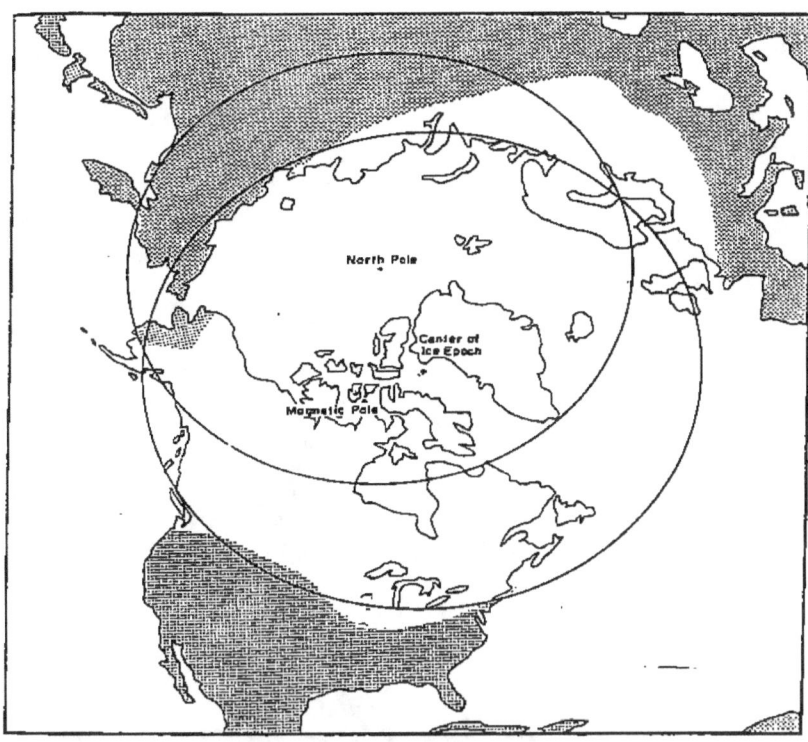

THE GEOGRAPHICAL LOCATION OF THE ICE EPOCH IN THE NORTHERN HEMISPHERE

with regard to ( 1 ) the Geographical Center of the Ice Age
( 2 ) the Geographical North Pole
( 3 ) the Magnetic North Pole

FIGURE 14

**Figure 3: Geographical Center of Glacial Ice**
(From the Book: *The Genesis Flood and the Ice Epoch*)

118 BIBLICAL FLOOD AND ICE EPOCH

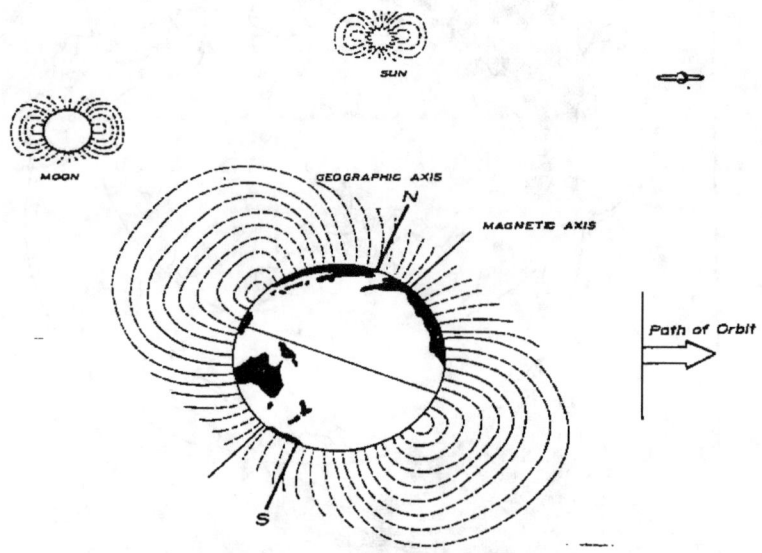

THE ASTROPHYSICAL LOCATION OF THE MAGNETIC FIELD WITH REGARD TO:

(1) Geographical Axis of Earth
(2) the Magnetic Poles &
(3) the Path of Orbit

AS VIEWED FROM 180° LONGITUDE

FIGURE 15

**Figure 4: Magnetic Fields of the Earth**

(From the Book: *The Genesis Flood and the Ice Epoch*)

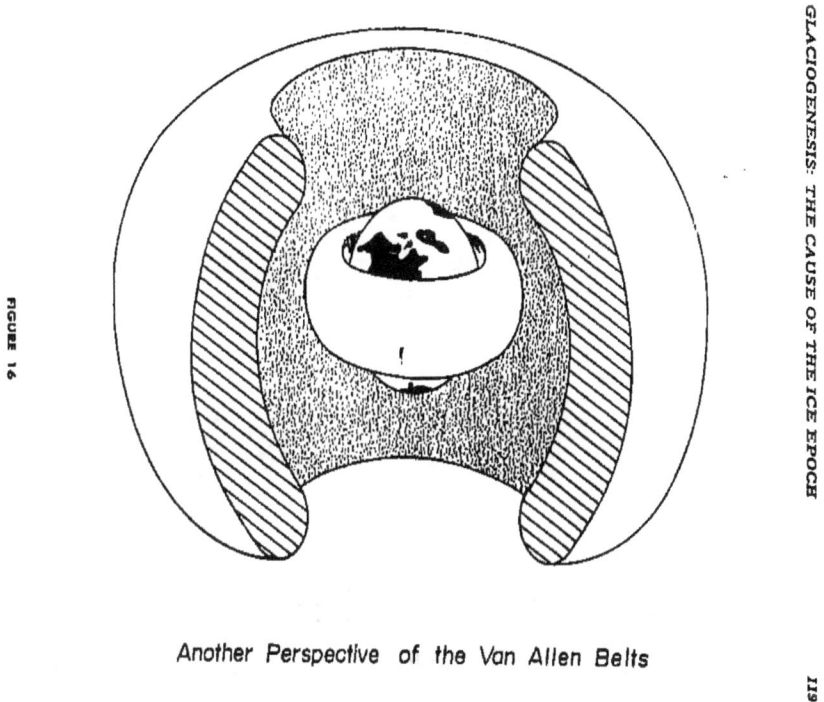

Another Perspective of the Van Allen Belts

**Figure 5: Van Allen Belts of the Earth**
(From the Book: *The Genesis Flood and the Ice Epoch*)

132                              BIBLICAL FLOOD AND ICE EPOCH

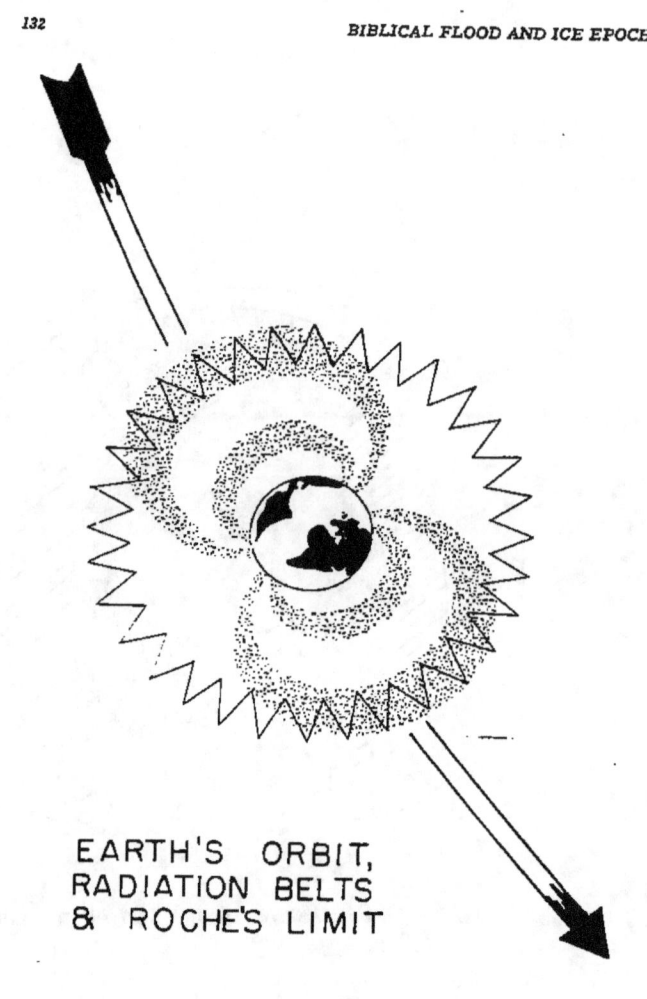

EARTH'S ORBIT,
RADIATION BELTS
& ROCHE'S LIMIT

FIGURE 18

**Figure** .. ............ ............ ..........

(From the Book: the Genesis Flood and the Ice Epoch)

GLACIOGENESIS: THE CAUSE OF THE ICE EPOCH 133

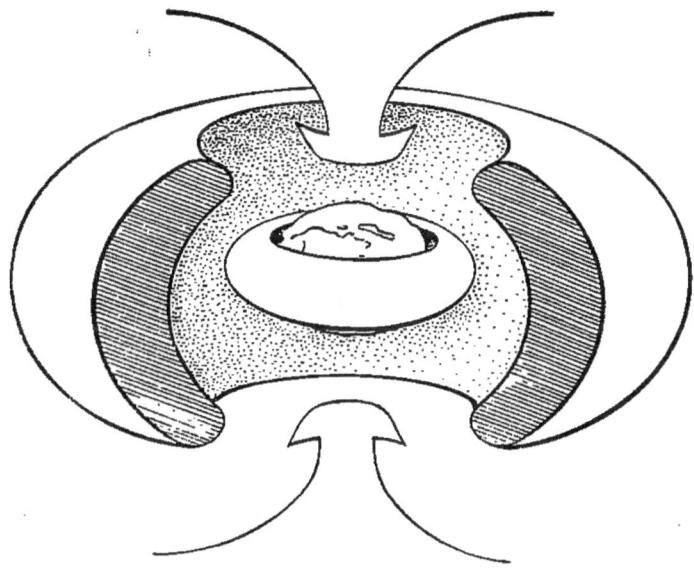

MANNER OF ICE DESCENT

FIGURE 19

**Figure 7: Manner of Ice Descent**
(From the Book: The Genesis Flood and the Ice Epoch)

M J Tiry

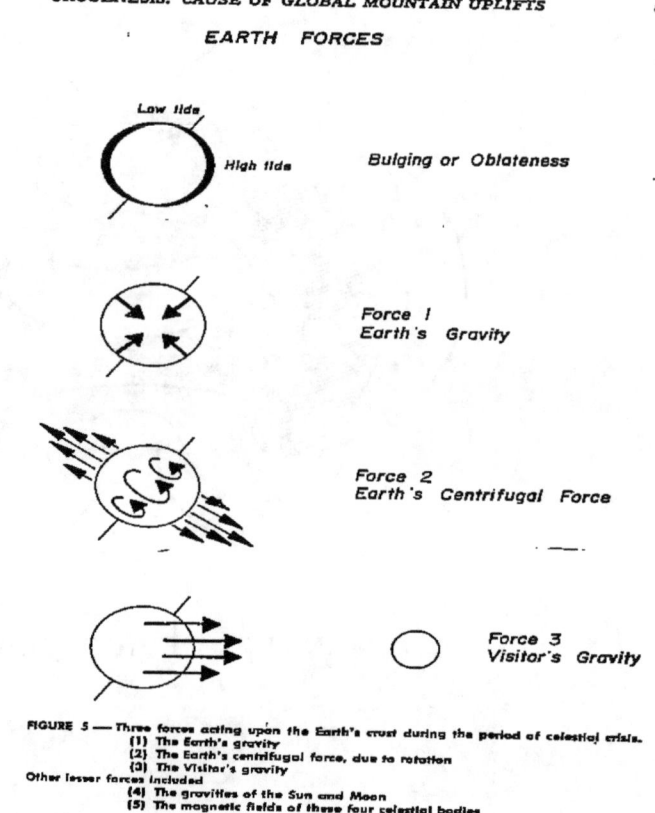

EARTH FORCES

Low tide

High tide

Bulging or Oblateness

Force 1
Earth's Gravity

Force 2  Earth's Centrifugal Force

Force 3
Visitor's Gravity

FIGURE 5 — Three forces acting upon the Earth's crust during the period of celestial crisis.
(1) The Earth's gravity
(2) The Earth's centrifugal force, due to rotation
(3) The Visitor's gravity
Other lesser forces included
(4) The gravities of the Sun and Moon
(5) The magnetic fields of these four celestial bodies

**Figure 8:  Forces Acting on the Earth**
(From the book: *The Genesis Flood and the Ice Epoch*)

128

# Genesis Annotations Chapter 11

## The Earth Divided

### Genesis 11:1-4 The whole world with one language

¹ And the whole earth was of one[a] language, and of one speech. ² And it came to pass, as they journeyed from the east, that they found a plain in the land of Shinar; and they dwelt there. ³ And they said one to another, Go to, let us make brick, and burn them throughly. And they had brick for stone, and slime had they for morter. ⁴ And they said, Go to, let us build us a city[c] and a tower, whose top *may reach* unto[d] heaven; and let us make us a name, lest[b] we be scattered abroad upon the face of the whole earth.

### [a] The Whole Earth One Language

Here in Chapter 11 we see that, up until the judgment of the confusion of tongues, the entire human race spoke the language that God gave to Adam. Noting that our Lord spoke to Saul of Tarsus from heaven in the Hebrew tongue, we are inclined to believe that Hebrew is not only the language of Heaven, but also the language that was given to Adam. In Zephaniah 3:9, we see that God will one day return to Israel (and to all the nations) a pure language "…that they may all call upon the name of the LORD, to serve him with one consent." It will be the language of Canaan, Egypt, and all of the earth in the Kingdom (Isa. 19:18) Note: The word Hebrew apparently comes from Eber, the father of Peleg.

### [b] Man Plans to Exclude God from His Creation and prevent Him from enjoying His Creator's Rights

The land of Shinar is Babylon, or modern day Iraq (Dan. 1:2). In Isaiah 11:11, this land is mentioned as one of the seven nations from which God re-gathers His people in the last day. This is the city of man and Satan's city. In Genesis 11, we see more than simply man defying God. What we see is man organizing for the express purpose of excluding God from the earth. Mankind in general here becomes a willing accomplice with Satan in an effort to deny God the right to enjoy His creation. He does this by means of a three-fold system:

1. Man is using a political system represented by the city of Genesis 11:4. Compare this with Genesis 10:9-10 where we see that Babel was built by Nimrod. With man exiting the arc, God commissioned them to "be fruitful and multiply…and bring forth abundantly in the earth." (Gen. 9:7). The stated purpose for the city was however to "…make us a name lest we be scattered abroad upon the face of the whole earth." (Gen. 11:4) This is Satan's city. We note that Satan's city comes first and then God's city. We find God's city (Jerusalem) appearing in the Bible in Genesis 14:18 where we find the fascinating Bible character Melchisedec being its king. There is a clear Bible principle that first comes the natural then the spiritual. For Example: First Cain, then Abel; First Ishmael, then Isaac; First Esau, then Jacob; First Saul then David; First the Antichrist, then Christ; First Natural birth, then New birth.

2. Man sets up an economic system – the welfare state. Nimrod was "…a mighty hunter before the LORD." (Gen. 10:9) God had cursed the ground for Adam's sake (i.e. for his good). Nimrod set out to provide protection from that curse by organizing humanity together. Here I believe that we see a commercial Babylon that would provide for people so that they did not have to provide for themselves. This would be a form of a welfare system – we could call it the welfare state.

3. Man also develops a religious system whose goal is to direct worship away from God and towards man. This system has been, is now, and will in the future be used by the devil to direct man's worship anywhere but to God. Ultimately, Satan will destroy this religious system (represented by the whore of Revelation Chapter 17) when it no longer serves his needs. At that time, he will direct all worship to himself (Rev. 17:16).

## [c] Man's City

Man, when he rejects God's will for him, must settle for that which is grossly inferior. Man's city was built of clay and slime. God's city (Rev. 21:11 ff) will be built of jasper stone, pearls, streets of gold, and foundations of jasper, chalcedony, emerald, sardonyx, Sardis, chrysulite, beryl, topaz, chrysoprasus, jacinth, and amethyst.

## [d] The Tower the Top from Which Man was to reach Heaven

There is something interesting about the top of the tower. They built a tower "…whose top may reach unto heaven…" Doubtless they knew that the tower could not possibly reach heaven physically. However, we understand that Satan is "the god of this world" (2 Cor. 4:4) and he is also today "the prince of the power of the air" (Eph. 2:2) the chief ruler of the invisible world of heaven. We understand from scripture that everything in heaven and everything in earth is created by Christ and for Christ (Col. 1:16-20). However we understand also from scripture that both realms are in the hands of a usurper. Satan also has many of the fallen angels to do his bidding (Eph. 6:12). His plan at Babel was apparently to rule both heaven and earth (through man) from that city. Note God's concern "…and this they began to do: and now nothing will be restrained from them which they have imagined to do…" (Gen. 11:6).

We ask ourselves here "What was it that they (man) had imagined to do?" What they imagined had something to do with the tower whose top may reach unto heaven. It is likely that the "all seeing eye" symbol found on the currency in use in the United States of America on the one dollar bill is the figure of that tower. This is all the more interesting when we consider that, when God sets up the kingdom of Matthew 25:34 that angels will ascend and descend from planet earth from the City of Jerusalem to the heavens where the church the Body of Christ will be reigning for Christ (2Cor. 5:1-4; 2Tim. 2:10-12). Our Lord foretells of this in John 1:51 though there was no way for anyone then to know what (or who) the communication from earth to heaven was to go to. Jacob, when he had the vision of angels ascending and descending on the ladder to heaven in Genesis 28:12, realized that he was at the spot on earth from which that communication would originate but could not possibly know about the mystery that Jesus finally revealed through Paul about a body of believers called the church the Body of Christ which would live and reign in the heavens (2Cor. 5:1-5; 2Tim. 2:10)).

Communication will be carried by angels from the earth to the heavens when that kingdom is established on earth. It is not until we come to Paul's epistles that we learn who it is that the communication is made with. In that portion of the inspired, inerrant Word of God that we call the Pauline epistles we find that there is a new creature called "the one new man" (Eph. 2:15) which will have a resurrection life "eternal in the heavens" (2 Cor. 5:1). This "one new man" is also called "the church which is Christ's body (Col. 1:18; Eph. 1:22; 5:23). This body of believers will "judge angels" (1Cor. 6:3) and "reign with Him [Christ] in the heavenly places. When that kingdom is set up, our Lord Jesus Christ will rule both heaven and earth from the New Jerusalem. I do believe that what we are seeing in Genesis 11 is Satan's plan to do the same thing. He would then be usurping what is created by Jesus and for Jesus and claiming it for himself. The bold faced audacity of the very thought is astounding to us – yet such is the nature of our adversary. There is little wonder why God took such strong action to counter this insurrection by man as man joined Satan in rebellion.

## Genesis 11:5-9 The Confusion of Tongues

[5] And the LORD came down to see the city and the tower, which the children of men builded. [6] And the LORD said, Behold, the people *is* one, and they have all one language; and this they begin to do: and now nothing will be restrained from them, which they have imagined to do. [7] Go to, let us go down, and there confound their language, that they may not understand one another's speech. [8] So the LORD scattered them abroad from thence upon the face of all the earth: and they left off to build the city. [9] Therefore is the name of it called Babel; because the LORD did there confound the language of all the earth: and from thence did the LORD scatter them abroad upon the face of all the earth.

## [e] Man's Plan Foiled

We see in verse 8 that man's plan was foiled. The scattering of the people is by no means the end of the city nor was it the end of man's efforts to bring in a world wide universal government that excludes God. The city of Babylon still has a future and will play a major role in end time events. Idolatry started there and the final form of idolatry will again be there. That religious, political, economic system is in the world today in a subtle hidden form. In Revelation 17:5, this system is called "Mystery Babylon the Great, the Mother of Harlots and Abominations of the Earth." It is "mystery Babylon" because only those who are regenerated and enlightened by the scriptures recognize it and understand it. In Revelation we see Satan destroy her. In Revelation 17, Religious Babylon is destroyed. In Revelation 18, Economic (Commercial) Babylon is destroyed. In Revelation 19, Political Babylon is destroyed. To see more on this subject go the book "*A Study in the Revelation*" (Tiry. 2023).

The final destruction of Babylon has not happened yet. In Isaiah 13:19 and 20, we see that once Babylon is destroyed, it will never be rebuilt. In 1Peter 5:13, we see that there were circumcision believers there in Babylon. Therefore this prophecy of Isaiah 13 has not been fulfilled yet. Revelation 14:8; 18:10, 20 state plainly "Babylon is that great city." That city will play a role in the end time events cited in the Book of the Revelation. In Zechariah 5:5-11 we see that this system will again be in the land of Shinar (Babylon). There are many interesting observations about this city in the scriptures:

- Just as the New Jerusalem (the city of God) has a husband (Rev. 19:7), so this city has a husband (Rev. 18:7) – that being Satan.
- The symbol of Babylon is the golden cup (Jer. 51:6).
- The city is the habitation of dragons, demons (Jer. 51:37).
- The city is the source of idolatry in the world (Jer. 50:38). The statement: "the wine of her fornication" is a reference to idolatry.
- The merchants of the earth will mourn her destruction (i.e. this city controls or will control the world's money system.)
- The city will be "the glory of kingdoms" before it is overthrown but will ultimately be as Sodom and Gomorrah (Isa. 13:19)
- Believers today are instructed to come out of that system and be separate (2Cor. 6:14-18, cf Isa. 52:11; Rev. 18:4)

## Genesis 11:10-26 - The Generations of Shem the Father of Abram

[10] These *are* the generations of Shem: Shem *was* an hundred years old, and begat Arphaxad two years after the flood: [11] And Shem lived after he begat Arphaxad five hundred years, and begat sons and daughters. [12] And Arphaxad lived five and thirty years, and begat Salah: [13] And Arphaxad lived after he begat Salah four hundred and three years, and begat sons and

daughters. [14] And Salah lived thirty years, and begat Eber: [15] And Salah lived after he begat Eber four hundred and three years, and begat sons and daughters. [16] And Eber lived four and thirty years, and begat Peleg: [17] And Eber lived after he begat Peleg four hundred and thirty years, and begat sons and daughters. [18] And Peleg lived thirty years, and begat Reu: [19] And Peleg lived after he begat Reu two hundred and nine years, and begat sons and daughters. [20] And Reu lived two and thirty years, and begat Serug: [21] And Reu lived after he begat Serug two hundred and seven years, and begat sons and daughters. [22] And Serug lived thirty years, and begat Nahor: [23] And Serug lived after he begat Nahor two hundred years, and begat sons and daughters. [24] And Nahor lived nine and twenty years, and begat Terah: [25] And Nahor lived after he begat Terah an hundred and nineteen years, and begat sons and daughters. [26] And Terah lived seventy years, and begat Abram, Nahor, and Haran.

## [f] Terah the Father of Abraham

Verse 26 takes us through Terah the father of Abram and through to the death of Terah in verse 32. That will be the end of God's dealings with the Gentiles until we come to the writings of Paul the apostles of the Gentiles in the Bible. Verse 27 then picks up with Abram who will become known as Abraham and the focus will then be on the seed of Abraham from here forward in the Bible.

## Genesis 11:27-32 The generations of Terah

[27] Now these *are* the generations of Terah: Terah begat Abram, Nahor, and Haran; and Haran begat Lot. [28] And Haran died before his father Terah in the land of his nativity, in Ur of the Chaldees. [29] And Abram and Nahor took them wives: the name of Abram's wife *was* Sarai; and the name of Nahor's wife, Milcah, the daughter of Haran, the father of Milcah, and the father of Iscah. [30] But Sarai was barren; she *had* no child. [31] And Terah took Abram his son, and Lot the son of Haran his son's son, and Sarai his daughter in law, his son Abram's wife; and they went forth with them from Ur of the Chaldees, to go into the land of Canaan; and they came unto Haran, and dwelt there. [32] And the days of Terah were two hundred and five years: and Terah died[f] in Haran.

Study Guide Questions on Chapter 11

1. What was man doing in verses 1 through 4 that caught God's attention?

2. What was it about the city and the tower that so displeased God?

3. We still find in the human race the issue that displeased God with the city and the tower. What is that social order that man was pursuing then called when we find it in society today?

4. Did man actually think that he could build that tower physically high enough to actually each heaven?

5. According to verse 6, God saw what it was that man imagined to do. What was it that man imagined to do?

6. What did God do to foil man's plan?

7. According to Romans 1:24, 1:26, and 1:28, God gave up on man at Babel in three ways. What were they?

8. When in human history did God again pick up with His dealings with the Gentiles?

# Genesis Annotations Chapter 12

## Father Abraham

### Genesis 12:1-3 (KJV)

[1] Now the LORD had said[a] unto Abram, Get thee out of thy country, and from thy kindred, and from thy father's house, unto a land that I will shew thee: [2] And I will make of thee a great nation, and I will bless thee, and make thy name great; and thou shalt be a blessing: [3] And I will bless them that bless thee, and curse him that curseth thee: and in thee shall all families of the earth be blessed.

### [a] The LORD begins to work with Abram

We first encountered Abram in Genesis 11:28 and 29 as the son of Terah. Here in Chapter 12 we start to see Abram come into full focus. In fact, from here on through the Bible from Genesis 12 to the middle of the Book of Acts in the New Testament scripture, the entire narrative is about this man and his seed. In Genesis 12:1 we see God telling Abram to leave his home and to go to a special land that He will show to him later. Matthew 25:34 tells us that what is happening here in Genesis 12 is what God had in mind from the foundation of the world. That plan would be the establishment a kingdom that would be set up on the earth through Abram's (Abraham's) multiplied seed. The Gentile nations were rejected by God in Genesis 11 and so in Genesis 12 God set one man aside from all of the rebellion to begin to work through him and his seed to form the nation that would be the heirs to that kingdom.

### Genesis 12:4-5 (KJV)

[4] So Abram departed, as the LORD had spoken unto him; and Lot went with him: and Abram *was* seventy and five years old when he departed out of Haran. [5] And Abram took[b] Sarai his wife, and Lot his brother's son, and all their substance that they had gathered, and the souls that they had gotten in Haran; and they went forth to go into the land of Canaan; and into the land of Canaan they came.

### [b] Incomplete obedience

What we witness here is incomplete obedience to God's command on Abram's part. He was supposed to leave his kindred and his father's house in verse 1 but here we see him taking his brother's son Lot with him. We will see that before God's provision for Abram can be fulfilled, this will have to be corrected – which we will see accomplished later in Genesis.

### Genesis 12:6-9 (KJV)

[6] And Abram passed through the land unto the place of Sichem, unto the plain of Moreh. And the[c] Canaanite *was* then in the land. [7] And the LORD appeared unto Abram, and said, Unto thy seed[d] will I give this land: and there builded he an altar[e] unto the LORD, who appeared unto him. [8] And he removed from thence unto a mountain on the east of Bethel, and pitched his tent, *having* Bethel on the west, and Hai on the east: and there he builded an altar unto the LORD, and called upon the name of the LORD. [9] And Abram journeyed, going on still toward the south.

## [c] The Canaanite and the Land

The Canaanite was then in the land. These were people who were put there by the adversary to block God's people from inheriting that land. It will eventually take the sword of Israel to dispossess them.

## [d] The start of the Abrahamic Covenant

"The Lord appeared unto Abram and said unto thy seed will I give this land." Here we see the first installment of what will be the Abrahamic Covenant in which God will give Abraham the land of Canaan for an everlasting possession. .

## [e] Abraham Grows in Faith

Abraham built an alter unto the LORD and there called upon the name of the LORD. We see the faith of Abraham grow as he matures in his faith. However we will also see lapses of that faith in several incidences in which we are amazed that a man who is viewed as a giant of such faith should fail. That should speak to our hearts as we too occasionally have such lapses of faith.

## Genesis 12:10-13 Abram in Egypt

[10] And there was a famine in the land: and Abram went down into Egypt to sojourn there; for the famine *was* grievous in the land. [11] And it came to pass, when he was come near to enter into Egypt, that he said unto Sarai his wife, Behold now, I know that thou *art* a fair woman to look upon: [12] Therefore it shall come to pass, when the Egyptians shall see thee, that they shall say, This *is* his wife: and they will kill[f] me, but they will save thee alive. [13] Say, I pray thee, thou *art* my sister: that it may be well with me for thy sake; and my soul shall live because of thee.

## [f] Abraham's Lapse of Faith

Here in verse 12 we see one of those lapses of faith. Had not God just recently promised that he would have a seed line that would inherit that land of Canaan? Yet here he is taking devious action to seek to prevent what he pictures might be action that men could take to interfere with God's plan.

## Genesis 12:14-20 The LORD over-rides Abram's lack of faith.

[14] And it came to pass, that, when Abram was come into Egypt, the Egyptians beheld the woman that she *was* very fair. [15] The princes also of Pharaoh saw her, and commended her before Pharaoh: and the woman was taken into Pharaoh's house. [16] And he entreated Abram well for her sake: and he had sheep, and oxen, and he asses, and menservants, and maidservants, and she asses, and camels. [17] And the LORD plagued[g] Pharaoh and his house with great plagues because of Sarai Abram's wife. [18] And Pharaoh called Abram, and said, What *is* this *that* thou hast done unto me? why didst thou not tell me that she *was* thy wife? [19] Why saidst thou, She *is* my sister? so I might have taken her to me to wife: now therefore behold thy wife, take *her*, and go thy way. [20] And Pharaoh commanded *his* men concerning him: and they sent him away, and his wife, and all that he had.

## [g] God overrules Abraham's lack of Faith

Here we find God overruling on Abraham's behalf in spite of his lack of faith. Here we find the heathen Pharaoh suffering because of Abraham's lack of faith and actually conducting himself more honorably than Abraham the man of faith. Interesting!

From Genesis chapter 12 through the rest of the Bible, the focus is on Abraham and his seed. The nation of Israel is the physical seed of Abraham and therefore that nation views Abraham as their father and therefore as themselves being the rightful seed of Abraham. The apostle Paul who is "…the apostle of the Gentiles…" (Romans 11:13) refers to Abraham as "…the father of us all…" (Romans 4:16). We will see in the Illustration by Robert Brook appended to this chapter how it is that the Gentile members of the church the Body of Christ can consider themselves as the seed of Abraham.

A brief overview of Abraham's life is as follows:
- In Genesis 12, the Lord instructs Abram to leave his country, his family, and his father's house and go to a country that He would show him.
- Abram visited the land and God there promised that He would give that land to Abram and to his seed. (Gen 12:7)
- Abram sojourned in Egypt for a time because of a famine in the Land of Canaan. (Gen. 12:10-20)
- God again appeared to Abraham after Abram separated from Lot (his brother's son) and again renewed the promised that He would give the land to Abram and his seed and told him that his seed would be as the dust of the ground for multitude. (Gen. 13:16)
- In Genesis 14 we find Abram rescuing Lot along with the inhabitants of Sodom and Gomorrah in a battle against four kings. Also in that chapter we see Abraham meeting Melchizedek and being blessed by Melchizedek and blessing Melchizedek in turn (cf Heb 7:4).
- In Genesis 15, Abram is again promised the seed (Gen 15:1-5). Abraham believed this promise and was reckoned righteous on the basis that he believed God (Gen 15:6). Here he is justified by faith without works and becomes a type of the believer today in the Dispensation of the Grace of God. (Rom. 4:1-5; Gal. 3: 6-9) I refer you to Appendix 24 for a comparison of how the uncircumcised Abram was justified by faith apart from works in comparison to how the circumcised Abraham was justified "…by works and not by faith only…" (James 2:20-22)
    - In verses 1-5 of Genesis 15 we see the promise of the son.
    - In verses 6-11 we see Abram justified by faith apart from works by simply believing a promise from God.
    - In verses 12-17 we see the horror of darkness vision (Prophetic of his seed's great trials of affliction).
    - In verses 18-21 we see again the promise of the land to Abram but with the further information that he would die before he possesses it. From this we understand that the basic promise of this Abrahamic Covenant is resurrection life.
- In Genesis 16: 1-6 we find Abram and Sarai trying to help God out in keeping His promise with the conception of Ishmael by Hager the Egyptian.
- In Genesis 17 we find again the promise of a multiplied seed.
    - Abram's name changed to Abraham. (Verses. 1-5)
    - Abraham receives the covenant of circumcision. (Verses 6-14)
    - The name Sarai's means "Bitter Woman.". Her name was changed to Sarah which means "Princess."
    - We find also the promised son named Isaac. (Verses 15-19)
    - The covenant will be with Isaac and not with Ishmael.
- In Genesis 18 we find Abraham negotiating with the Lord regarding Sodom in order to save Lot.
- In Genesis 19 we find the destruction of Sodom and Gomorrah.
- In Genesis 20 we see Abraham's second lapse of faith in the issue regarding Sarah and Abimelech.

- In Genesis 21 we see the birth of Isaac to Abraham when Abraham is 100 years old. Ishmael is sent away. There is also an issue with Abimelech over a well.
- In Genesis 22 we come to the test that God makes on Abraham's faith with God asking Abraham to offer up the promised son in sacrifice. Here Isaac (the promised seed) is a type of Christ the real promised seed. Abraham passes the test and God stops him from going through with it. Here we find a type of the faith that Israel (The multiplied seed of Abraham) will have to have for justification. James, in James 2:27, writing to the twelve tribes of Israel cites this experience in Abraham's life as the example of how an Israelite in the tribulation will be justified – by faith that will be demonstrated by works. The apostle Paul however, takes us Gentiles living in the dispensation of grace back to the experience of the uncircumcised Abram who was justified by faith without works (Genesis 15:6).

## Abraham as the Father of us All

There are actually three different seed of Abraham that can be identified in the Bible. One of course is the physical descendents of Abraham through whom the promises to Abraham will be fulfilled. God made a promise to Abraham in Genesis 15:1-5 that Abraham would have a son. That seed would be the nation of Israel.

Another seed line would be the Gentile believers who are members of the Body of Christ. Today, all believers whether they be Jew or Gentile are baptized into an eternal union with Jesus Christ when they believe in the redeeming work of Christ on Calvary (Rom 6:1-4; Gal 3:27) and are thereby the seed of Abraham (Gal 3:29) because they are joined spiritually to the Lord Jesus Christ the seed of Abraham. The real seed though is Christ (Gal 3:16). The only way either Israel or the Body of Christ can be the seed of Abraham is that individual members of each become the seed of Abraham by faith.

See Appendix 24 in the study of how the Gentile members of the Body of Christ are today the seed of Abraham. Note to the reader: This illustration is a presentation in graphical form of the doctrine of the fatherhood of Abraham as the father of all believers. The doctrine is presented in Galatians Chapter 3. Appendix 24 presents an expository study of Galatians Chapter 3 with annotations to help the reader understand this key point of doctrine.

Illustration by Robert Brook of Abraham as "The Father of Us All" in Romans 4:16
Robert Brook is now home with the Lord. This illustration is used by permission of his family.

| | |
|---|---|
| The Uncircumcised Abram (Deut 26:5) | The Circumcised Abraham |
| Only child begotten vs. Ishmael (Gen 16:15) | The Father of Nations (Gen 17:4) |
| Gospel preached to Abraham (Gen 12:1-3, 15:1-5) | Isaac – Only Son (Gen 22:2) |
| | Sons of Ketorah (Gen 25:1-6) |

Genesis 17:15  Abram

99 | Abraham + 76 years = 175

Circumcised

**Gentile**
Seed as the dust of the earth (Gen 13:16)

**Hebrew**
Seed of the dust of the earth (Gen 28:14)

(b) They which are of faith the same are the children of Abraham (Gal 3:7)

Spiritual seed by faith

Type of the Believer today (Gal 3:8-9)

(c) Physical seed by promise

As the stars of Heaven (Gen 15:5)

Isaac | Jacob

(f) Gal 3:16
The Seed
(singular)

Gal 3:16
The Seed
(plural)

As the Sand of the Sea shore (Gen 22:17 cf 32:12)

Unbelieving Israel
(Mat 21:43, Rom 9:6-7)

(a) The Body of Christ

Christ

(d) Believers of Israel (Luke 12:32)

Father of us all
(Rom 4:16)

141

Study Guide Questions on Chapter 12

1.  In Verse 1, God told Abram to leave his kindred and his father's house. Did he entirely obey that command?

2.  What did God promise to Abram and his seed?

3.  In Romans 11:3, Paul tells us that Abraham is the father of us all. Is he your father?

4.  In what way is he your father? Are you a physical descendent of Abraham?

# Appendix 1: A Framework for Dating Creation

## Based on Information in the Bible

The Bible provides a framework for the dating of the creation of man in the earth. This framework enables us to use the information in the Bible to make an accurate time line of human history in the earth.

1. Genesis chapter five contains chronological data from the time of Adam to the time of the Flood of Noah.
2. Genesis chapter eleven presents the chronology from the Flood of Noah to Abraham.
3. The historical books of the Old Testament contain the chronological data of the nation of Israel from Abraham to the Babylonian captivity.
4. The chronology of the captivity and the restoration is obtained from the prophetic books of Isaiah, Jeremiah, and Daniel and the historical books (post captivity) of Ezra and Nehemiah
5. The chronology of the inter-testament period is presented clearly in the "Seventy Weeks of Daniel" prophesied in Daniel chapter 9.

Computed Dates for the Creation of Man
- Archbishop James Ussher (1581- 1656) set the date at 4004 B.C.
- Isaac Newton "The Chronology of Ancient Kingdoms Amended" agreed with Ussher implicitly.
- Others:

| | |
|---|---|
| Jewish | 3760 B.C. |
| Josephus | 5555B.C. |
| Kepler | 3993 B.C. |
| Melanchton | 3964 B.C. |
| Luther | 3961 B.C. |
| Lightfoot | 3960 B.C. |
| Holes | 5402 B.C. |
| Ployfair | 4005 B.C. |
| Lipman | 3916 B.C. |
| Floyd Nolen Jones | 4004 BC |

Floyd Nolen Jones presents a scholarly Chronology of the Old Testament in his web site floydnolenjonesministries.com and in his book by the same title. A Summary of his rationale for the use of 4004 BC for the creation of Adam is presented in the Preface to his book.

## Timeline of Genesis

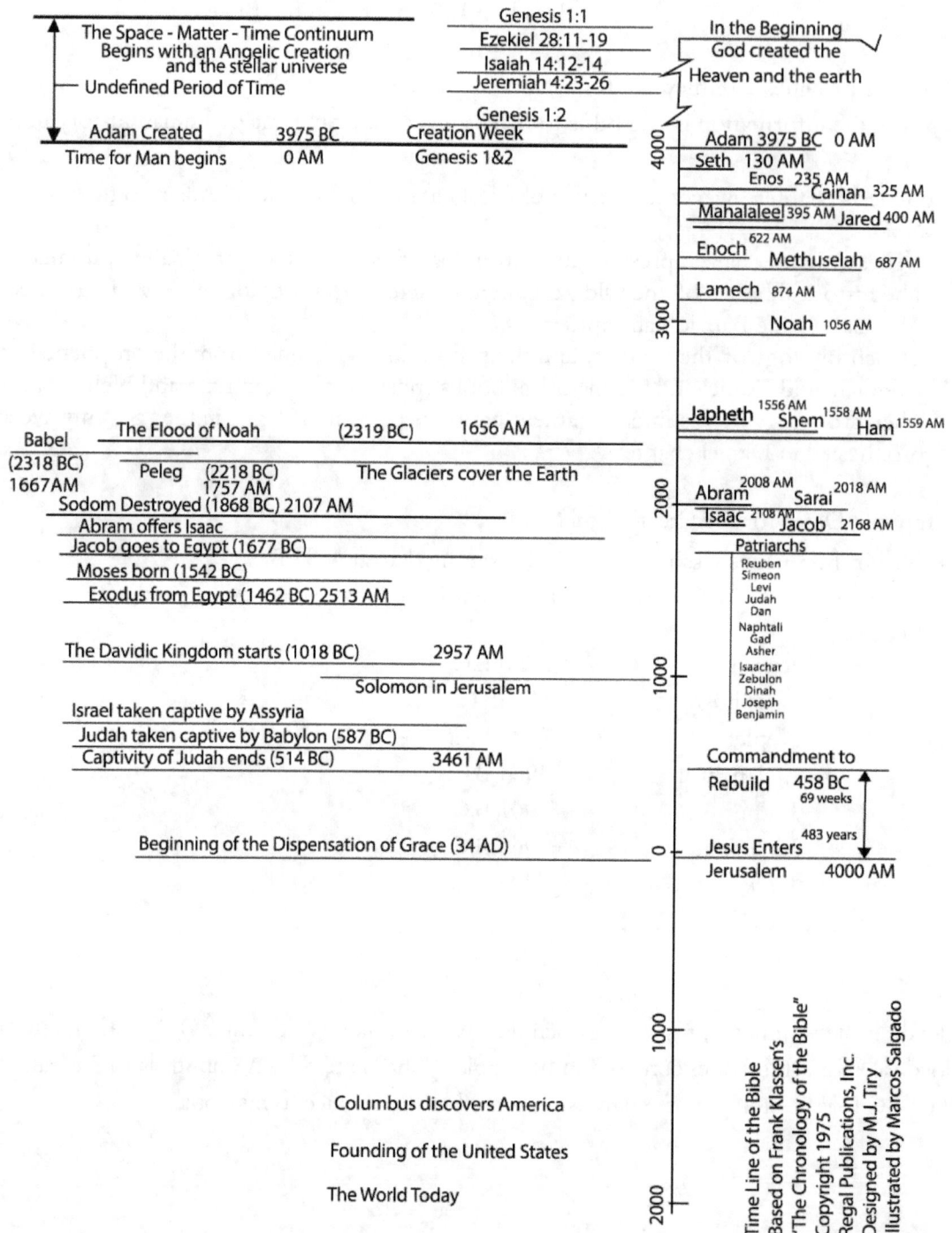

Note: The date for the creation of Adam used in this chart is taken as 3975 BC. This date is from Frank R. Klassen in his book *The Chronology of the Bible*. I have followed his logic and am in agreement with it. The rationale for the computed date for the creation of Adam used by Ussher, Isaac Newton, and Floyd Nolen Jones of 4004 is compelling as well. To convert from the BC dates in Figure 8 to the dates of Ussher or Jones, simply add 29 years to the BC dates in the chart. The AM dates in the chart will remain the same as theirs.

**Figure 9: Timeline of Genesis and God's work in the Earth**

# Appendix 2: "And God Said..."

This expression ("And God said...") is used ten times in Genesis Chapter 1. This simple statement exemplifies the power of the Word of God. God called the universe into existence by the power of His word. Our Lord Jesus Christ is called "the Word" in connection with Him being the one who created everything that has been created (John 1:1-3). In the beginning of time, the entire space-matter-time continuum came into existence at the command of the one who is called the Word. In John 1:14 we see that the one who is the eternal Word who made everything that has been made was made flesh entered into the humanity that he created. As we go into creation week, we see the word "let..." used as the command for acts of creation by the Word to cause it to happen. This statement "Let..." carries with it the thought that something that was planned in intricate detail was given permission to be and thus was brought into existence by God. We see the power of the Word in Hebrews 11:3 and how our faith enables us to understand how that can be. "By faith we understand that the worlds were framed by the <u>word</u> of God..."

"And God said:"

| | | |
|---|---|---|
| Day 1 | Gen. 1:2 | "Let there be light" |
| Day 2 | Gen. 1:6 | "Let there be a firmament in the midst of the waters..." |
| | Gen. 1:9 | "Let the waters under the heaven be gathered together in one place..." |
| Day 3 | Gen. 1:11 | "Let the earth bring forth grass, the herb yielding seed, and the fruit tree..." |
| Day 4 | Gen. 1:14 | "Let there be lights in the firmament of the heaven to divide the day from..." |
| Day 5 | Gen. 1:20 | "Let the waters bring forth abundantly the moving creature that hath life..." |
| Day 6 | Gen. 1:24 | "Let the earth bring forth the living creature after his kind..." |
| | Gen. 1:26 | "Let us make man in our image, after our likeness: and let him have dominion..." |
| | Gen. 1:28 | "Be fruitful and multiply and replenish the earth, and subdue it and have dominion over..." |
| | Gen 1:29 | "Behold, I have given you every herb bearing seed... and every tree, in which is the fruit of a tree yielding seed ..." |

## Appendix 3: Five Attacks on the Word of God

In Genesis Chapter 3 we see a series of attacks on the spoken Word of God by Satan the adversary. As we follow the series of five steps that the adversary took to bring about the fall of man, we learn how Satan attacks the written Word of God today. As we follow the series of steps, we see that gullible man was complicit in the deception by failing to stand on the truth of the Word of God..

> [1] Now the serpent was more subtil than any beast of the field which the LORD God had made. And he said unto the woman, Yea, hath God said, Ye shall not eat of every tree of the garden?[2] And the woman said unto the serpent, We may eat of the fruit of the trees of the garden: [3] But of the fruit of the tree which *is* in the midst of the garden, God hath said, Ye shall not eat of it, neither shall ye touch it, lest ye die. [4] And the serpent said unto the woman, Ye shall not surely die: [5] For God doth know that in the day ye eat thereof, then your eyes shall be opened, and ye shall be as gods, knowing good and evil.

**Genesis 3:1-5 (KJV)**

In step 1, He questions the Truth of it – "Yea, hath God said, Ye shall not eat of every tree of the garden?" We today face the same line of questioning: Did God really say that?

In step 2, the woman became complicit in her own deception by adding to the word saying "Neither shall ye touch it."

In step 3, she detracts from the Word by watering it down saying "lest ye die…" instead of "Thou shalt surly die." It is a common ploy of the traditions of men today to soften the consequences of violating the commands of God or to detract from Word today by excluding parts of it. Men today make a liberal interpretation of the Word so that it does not carry the authority it is designed by God to have.

In step 4 we see Satan make a flat out denial of the truth of the Word of God saying "Thou shalt not surly die." Men do that today by denying the fact that the Word of God is just that – the words that God spoke.

There is a fifth step here worthy of careful consideration in the seductive process. Not only did Satan succeed in his denial of the truth but he convinced the woman that God was denying her some pleasure that she could justly have.

## Appendix 4: Dispensations and Institutions of God

A Dispensation defines the state of man during a defined period of human history and how God interacts with man while man was in that particular state. The chart below shows seven dispensations that describe God's dealings with man during successive periods of time through history. The dispensations that impact God's interactions with man include: Innocence, Promise, Law, Grace, the Kingdom, the Fullness of Times, and Before the World began. See Table 6 in Chapter 3.

| Time Past | | | | But Now | Ages to Come | |
|---|---|---|---|---|---|---|
| | Adam in Eden | Adam to Moses | Moses | Paul | Christ | |
| Before the World Began — God made a promise to Himself of eternal life (Titus 1:2) | Innocence — Man walked and talked with God | Promise (Mat 25:34) — Adam · Adam (sin entered) · Noah · Shem · Abraham · Isaac · Jacob · 12 Tribes · Moses | Law — Circumcision Uncircumcision (Ephesians 2:12) | Grace — One new man Jew and Gentile saved in one body | Kingdom — Christ reigns on earth over Israel and Israel a blessing to the nations | Dispensation of the Fullness of Times (Eph 1:10) |
| The Angelic Creation | Adam was to Reign | Death Reigned (Rom 5:14) | Sin Reigned (Rom 5:12, 21) | Grace Reigns Rom 5:20, 21 | Righteousness Reigns | All things reconciled to God (Col 1:20) |

Genesis 3:15 begins the Dispensation of Promise.
The Dispensation of Innocence ended with Genesis 3:6-7.

| | |
|---|---|
| Romans 5:12 | By one man sin entered the world and death by sin. |
| | Death passed upon all men for all have sinned [in Adam]. |
| | Sin was in the world. |
| | Sin is not imputed when there is no law. |
| | Nevertheless, death reigned from Adam to Moses. |
| Gal 3:19 | The law was added to the promise. |
| | The promise was before the law. |
| | The law was added to the promise; it did not replace the promise. |
| | The law was interrupted by grace. |
| | Law will end with the New Covenant. |
| | The promise is fulfilled in the New Covenant, but we have the benefit of it under grace. |
| By Adam ☐ | Entrance of sin. |
| By Moses ☐ | Entrance of the knowledge of sin. |
| By Paul ☐ | The forgiveness of sin. |

Human government is an institution, not a dispensation.
Conscience is really part of volition and is also an institution, not a dispensation.
From Adam onward, promise was the issue (Matthew 25:34).

# Appendix 5: Genesis and Geology

The phenomenon that we observe in the earth's surface can be explained by the events of Genesis. Geology is the study of the earth's surface. It is a fascinating study for the believer because it testifies to the truth of the Bible. However, as in many other fields of science (science falsely so called – 1Tim. 6:20), the field of geology as presented in secular textbooks is biased toward agnosticism and atheism. They present an inaccurate view of the facts so as to accommodate the world's natural reluctance to accept any testimony to the truth and reliability of the Bible.

Let's consider the study of geology in a very general way as to the reliability of the Genesis account relative to what we can observe in the real world. The study of geology can be divided into two categories; bedrock geology and glacial geology. Bedrock geology bears on the study of the rock formations that underlay the soil mantle. Bedrock formations are of three types: igneous, metaphoric, and sedimentary.

1.  The igneous rock formations are the basic rock strata that underlay all others. They were formed when the earth was originally formed. They are the crystalline rocks such as basalts and granites formed when liquid magmas (molten rock from deep within the earth) cool. Such magmas never exist long on the earth.

2.  Sedimentary rocks are rock formations that are formed as sands, silts, clays and gravels are dropped out of either standing or moving water with a cementing agent. These rock formations cover most of the earth's surface and contain the fossils. As we study the different types of sedimentary rocks, we can picture the sequential deposition of progressively finer particles from course gravels to the finest clays. If we were to take a beaker, put all of the different sized particles in it with sufficient water to immerse it all, and then mix it and allow it to settle, we would see a column of material form that would essentially duplicate what people call the geologic column.

3.  Let's consider the sedimentary rock formations that we see in the earth and compare them with what we would see in that beaker:

-   Conglomerates are composed of cemented gravels and boulders with interstitial sands and pebbles. Very strong currents would be required to produce conglomerates. The first stages of the universal flood would have provided the violent currents needed to produce the conglomerates. There are no conglomerates being formed in the earth today.

-   Sandstone consists of sand particles of various sizes deposited out of running water or slowly moving water along with a cementing agent.

-   Shale is like sandstone but consist of silt and clay particles. Mud held in suspension by turbulent water until deposit in quieter water would produce shale.

-   Limestone and Dolomite are chemical sediments. Limestone is cemented with calcium carbonate while dolomite is cemented with calcium magnesium carbonate. Marine organisms secrete calcium carbonate compounds. Therefore, limestone and dolomite could be formed in the earth today (e.g. coral reefs). However, limestone and dolomite sediments as found in the geologic column are not being formed today. The calcium and magnesium rich floodwater could have formed the vast deposits very quickly.

4.  Metamorphic rocks are igneous and sedimentary rocks that have been structurally altered by subjection to heat and pressure.

5.  Coal is vegetation that has been altered by heat and pressure to produce an energy rich substance that can by burned to release its stored energy. It is found inter-bedded with sediments. Most coal deposits consist of trees that had been deposited in random order as if strewn by floodwater. We can understand the mechanism for the depositing of trees in this manner by considering the effect of trees floating in bays and coves as the floodwater recedes. The incoming tides would bring in a fresh supply of floating trees and the outgoing tide would leave a layer of sediment to cover it. Eventually, there would be sufficient heat and pressure to form coal. Coal can be made in the laboratory today by subjecting wood to heat and pressure.

6.  Oil and petroleum is fossilized organic compounds formed by heat and pressure under aqueous conditions. The organic matter would have come from fish and animal matter trapped in the sediments of the floodwater. Oil is found in the sediment deposits that were laid down in the floodwater. Basically our society today is powered by the organisms that were in "the world that perished" in the flood of Noah.

Glacial geology is a special subject that is covered in the annotations on chapter 10 of this study. See notes [c] and [d] in the annotations of that chapter regarding the days of Peleg.

See the figures below with the captions on some of the many evidences of the geological history of our earth as it is recorded in the Book of Genesis.

**Figure 10: Sedimentary Rock Formation**

**Figure 11:  Eroded Sedimentary Rock Formation**

Note the erosion producing the geology of the Coulee Region of Southwest Wisconsin, Southeast Minnesota, and Northeast Iowa. These would be eroded out of the sedimentary deposits from the flood of Noah's day.

**Figure 12: View of Rock Fence Made of "Fieldstone"**

This is rock material that had been ripped from the various rock formations over which the glacier passed. The various rock fragments were then rolled in the glacier and were rolled to rounded smooth shapes. The degree of rounding and smoothing of the rock was directly proportional to how far the rock material had been transported by the glacier. Looking at this picture, note that there are field stones that came from about 10 different rock formations. This particular picture was taken in Sun Prairie, Wisconsin. There is much of this glacial till material in the Kettle Moraine area of Southeast Wisconsin. The kettle holes of Southeast Wisconsin were formed as ice was trapped in glacier deposits of till material. As the ice melted, it left depressions called "kettle holes." Other glacial features, such as, drumlins, caimes, and eskers are abundant in that area as well. This is an indication of how massive the glacier was.

**Figure 23:  Glacial Gravel**

This photo shows gravel that had been ground to a small size by the action of the glacier. This photo is taken of a landscape design but the gravel material shown had been taken from a glacial out wash plane. The fine material "rock dust" that had been ground off of the rock fragments become a part of the silt sized soil particles that make up much of the soils of Wisconsin. Glacial till is a mix of everything that a glacier picked up as it traversed a landscape (e.g. rocks, sand, silts, clays, etc.). Note that this photo shows a mix of different rock materials (igneous rock, sandstone, shale, and limestone) indicating that the glacier that transported the material had passed over rock formations composed of theses different materials. Deposits of sand and gravel sized particles usually are found in river beds in glaciated areas. The Chippewa River valley is a deposit of glacial outwash sand. The melt wasters from the glacier deposited the material in the river valley as the glacier receded.

# Appendix 6: Genesis and the False Theories on Origins

Genesis 1:1 refutes all of the false philosophies regarding the origin and purpose for creation.

It refutes:

- Atheism because it is clear to rational thought that nothing could not have created everything. A purposeful God created it all.
- Pantheism because God exists totally outside of His creation.
- Polytheism because One God (the triune Godhead) created all things and interacts with it.
- Materialism because all space, matter, and time had a beginning.
- Dualism because the triune God was alone when He created it
- Humanism because God and not man is sovereign over His creation and is the ultimate authority over it.
- Evolution because God created all life so that it can reproduce "after its kind."

And most importantly, God left us a testimony of Himself in the Bible -- the Word of God.

### Appendix 7: Henry Grube's Outline of Genesis

Henry Grube was a Bible Teacher at Bob Jones University. He laid out his presentation of Genesis according to the breakdown below by chapters. I present it here because of its simplicity.

| | | |
|---|---|---|
| **Gen. 1 & 2 Creation** | **Four Events:** | **1. Creation** |
| **Gen. 3 & 4 Corruption** | | **2. Fall** |
| **Gen. 5 - 10 Catastrophe** | | **3. The Flood** |
| **Gen. 11    Confusion** | | **4. Babel** |
| **2000 years of history** | | |
| | **Four Institutions:** | **1. Volition** |
| | | **2. Marriage** |
| | | **3. Family** |
| | | **4. Sovereign Nations** |
| *Gen 12 – 50 Chosen Race* | *Four Men:* | *1. Abraham* |
| | | **2. Isaac** |
| | | **3. Jacob** |
| | | **4. Joseph** |
| **350 years of history** | | |

# Appendix 8: How Do We Account for the Existence of the Universe?

The laws of science when applied to all of the theories put forth by men show to the reasoning mind that only one explanation gives a satisfactory answer to the question of origin. Theories on the explanation of the existence of the universe fall into three categories by rational thought.

They are:

1. The universe has a mind of its own and created itself.

2. The universe always existed from eternity past.

3. The universe had a supernatural creation by an omnipotent God who created it purposefully out of nothing.

Consider these three theories in light of the laws of science;

The First Law of Science:

"Matter and energy are interchangeable but the sum total is fixed – it can not be created nor destroyed."

This would refute the theory that the universe created itself or came into existence apart from the laws that presently govern it.

The Second Law of Science:

*"In every energy transformation process, a certain amount of energy goes into its waste form of heat."*
This would refute the theory that the universe always existed. If it did, then, all of the energy would have gone into its waste form of heat and the universe would have died a heat death. Since this has not happened, we conclude that the universe is not infinitely old and therefore had a beginning at a point in time in the past.

This leaves us with only one possibility; that of supernatural creation by an omnipotent creator. The Book of Genesis gives us a satisfactory and irrefutable account of God acting to create the heaven and the earth.

## Appendix 9: How Not to Date a Rock

The apostle Paul give advice to the young man Timothy in 1Timothy 6:20-21 saying "O Timothy, keep that which is committed to thy trust, avoiding profane *and* vain babblings, and oppositions of science falsely so called: Which some professing have erred concerning the faith." There is much profane (worldly) and vane (empty) babblings and the oppositions of science falsely so called in the world. There is true science which establishes the truth of the Genesis account of creation and there is a false science that is crafted by men who seek to deny the truth of a purposeful God bringing about His creation. This false science uses what we call syllogistic reasoning in which the rocks are used to date the fossils that are contained in them and the fossils are used to date the rocks that contain them. However, the whole system was worked out to give the impression of life having evolved over a period of millions and billions of years and to have done so without the input of a creator. Listed below is a list of guidelines of how not to date a rock:

1. Rocks are not dated by their appearance. "Old" rocks do not necessarily look old. "Young" rocks do not necessarily look young.
2. Rocks are not dated by their makeup. Shales, granites, limestones, conglomerates, sandstones, etc. may be found in all "age" groupings of rocks.
3. Rocks are not dated by their mineralogical contents. Minerals, metallic ores, and even petroleum may be found in rocks of practically any "age".
4. Rocks are not dated by their adjacent rocks. Rocks of supposedly "older" ages may be found resting conformably on top of supposedly "younger" rocks.
5. Rocks are not dated by their structural features. Faults, folds, etc. may be found in rocks of practically any "age."
6. Rocks can not be dated radio-metrically. The geologic column and it's supposed "ages" were worked out based on their fossil content and the theory of evolution long before radioactive dating was thought of.

## Appendix 10: Man: Created in the Image and Likeness of God

"And God said, Let us make man in our image, after our likeness: and let them have dominion over the fish of the sea, and over the fowl of the air, and over the cattle, and over all the earth, and over every creeping thing that creepeth upon the earth. 27 So God created man in his *own* image, in the image of God created he him; male and female created he them." (Gen 1:26-27)

**The Image of God** – The Trinity of the Godhead

According to this passage in Genesis, God created man in His image. If we are to understand man and man's makeup, we need to look at God and learn about Him. Our only way to know about God is in the written Word of God -- the Bible.  As we read and study the Word of God, we find that there is one God (Deuteronomy 6:4) but we find also that there are three persons who are called God in the Bible. In this passage in Genesis, we see that God [singular] said "let us [plural] make man [singular] in our [plural] image [singular]…" In Isiah 45:5-7, God said "I *am* the LORD, and *there is* none else, *there is* no God beside me: I girded thee, though thou hast not known me: That they may know from the rising of the sun, and from the west, that *there is* none beside me. I *am* the LORD, and *there is* none else. I form the light, and create darkness: I make peace, and create evil: I the LORD do all these *things*."

The Bible says that there is one God and yet in the Bible there are three persons in the Bible called God. Each has their own personality. Each operate independently of each other but when they act, they do so in perfect harmony of purpose with each other. God is called "the Father" in the Bible (Romans 1:7). He is called the Father because He is the Father of our Lord Jesus Christ (Romans 15:6). He has a Son (Proverbs 30:4) who He (the Father) also calls God: "But unto the Son *he saith*, Thy throne, O God, *is* for ever and ever: a sceptre of righteousness *is* the sceptre of thy kingdom. Thou hast loved righteousness, and hated iniquity; therefore God, *even* thy God, hath anointed thee with the oil of gladness above thy fellows.  And, Thou, Lord, in the beginning hast laid the foundation of the earth; and the heavens are the works of thine hands:" (Hebrews 1:8 - 10).

The one who the Father calls the Son is the eternal Word from eternity past. We find the reference to the eternal Word in John 1:1-4 where we read "In the beginning was the Word, and the Word was with God, and the Word was God. The same was in the beginning with God. All things were made by him; and without him was not any thing made that was made. In him was life; and the life was the light of men." In this passage we find two persons called God. In Acts 5:4 we meet one called the Holy Ghost and find that he too is called God. This brings us to the concept of the Trinity. People who oppose the doctrine of the trinity readily point out that the word "Trinity" is not even in the Bible. That is true but the doctrine is clearly there. The Bible word for the Trinity of God is the word "Godhead" as we see it in Acts 17:29; Romans 1:20; and Colossians 2:9.

Man as a Three Part Creature:
The three part makeup of man as body, soul, and spirit is seen in 1Thessalonians 5:23 "And the very God of peace sanctify you wholly; and *I pray God* your whole spirit and soul and body be preserved blameless unto the coming of our Lord Jesus Christ."

We members of this human race are three part creatures -- each part of which has a mentality:
>    In our bodies, we each have a mentality that the Bible calls the **flesh.**
>    In our souls, we each have a mentality called our **heart.**
>    In our spirits, we each have a mentality called our **mind.**

157

## Our Human Spirit

The human spirit is given to us as God's creatures to relate to Him. It gives us a God consciousness. It enables us:

- To relate to God (Rom. 8:16).
- To receive information from God through the Word of God (1Cor. 2:16).
- To understand the things of God (1Cor. 2:11-12).

What the Bible calls the "Mind" (as we find it in the Scriptures) is the mental content of the human spirit. It is that part of our makeup that enables the Holy Spirit of God to communicate with us.

- In the unsaved person the mind is "reprobate" (Rom. 1:28).
- In the saved person, the mind serves the law of God (Rom. 7:25).
- The mind of the saved can be carnal (Rom. 8:27).
- Scripture is the means of renewing the mind (Rom 12:2; Eph. 4:23).
- The spiritual person has the mind of the Lord – having acquired it from a study of the Word of God (1Cor. 2:16).

## Appendix 11: Our Human Soul

While the human spirit give us humans a God consciousness, the human soul gives us self consciousness – our identity as a unique person. The soul of you is the real you and will be you through all of eternity. The Pauline passage in 1 Corinthians 6:20 is instructive on the role of our soul in our human existence: "For ye are bought with a price: therefore glorify God in your body, and in your spirit, which are God's." This passage informs us that we are a soul which (who) has (possesses) a body and who also has a spirit and you (the soul that is you) decides what you will do with both. The soul is the decision making part of our human makeup. It gives us:

- our identity
- our personality
- our emotional make-up.

The soul has a mentality that the Bible calls the Heart. It is the heart that believes (Rom. 10:9 & 10). The "Heart" as we find it in scripture:

Manifests itself in our speech (Matt. 12:34).

Is the sum total of our convictions (Prov. 23:7).

The beliefs and convictions of the heart produce our emotions.

The heart can:       grieve (Gen. 6:6)

communicate to others (Gen. 17:17)

be glad (Ex. 4:14)

be hardened (Ex. 4:21)

be sincere (Num. 21:39)

be discouraged (Deut. 1:26)

it can fear (Deut. 7:17)

it is the center for love in our lives (Deut 6:5)

But, the emotions of the heart are not the primary driver of our thinking but rather are designed by God to follow our thinking. Therefore, the heart can be deceived (Deut 11:16).

# Appendix 12: Our Human Body

The soul and spirit can exist and function without the body as we see in the account of the rich man and Lazarus in Luke 16:19 through 31. Physical life is defined in the Bible as the union of the soul and spirit with the body (Gen. 2:7). Physical death is the departure of the soul from the body (Gen. 35:18). In 2Corinthians 5:1-5 we see that this present body is called a tabernacle – a temporary dwelling for the soul. The resurrection body that we will one day have is called a "house not made with hands" in that passage. For us who are saved (justified) persons living in the dispensation of grace, that permanent body will be "eternal in the heavens."

The human body gives us world consciousness. It is the means whereby we live out the will and the desires of the soul and spirit in the physical universe. The human brain is not the thinking or the decision making part of our being but is only the medium that enables us to live out the thoughts and the intents of the heart and the mind while we are alive in this world.

What we do in and through our bodies tells those who observe us what is in our hearts and minds. Godliness starts with a sound mind – a spirit that is tuned into the things of God that we find in the Word of God the Bible..

## The Likeness of God

This is the original Christ likeness that was in man when he was first created (Luke 3:35).
This likeness was lost in the fall.

Man is still in the image of God (Gen. 9:6), but man is also in the likeness of fallen Adam (Gen. 5:1-3).

However, god-like-ness (godliness) is still available to man but it can only be found "In Christ" (1Tim. 3:16; 2:10; 4:7&8; 6:3-5; Tit.1:11). God is in the process of conforming believers into the image of Christ as they humbly yield to the working of God the Holy Spirit working through the Word of God (Romans 8:28).

## Appendix 13: Name of the Patriarchs

The names of each of the patriarchs have a meaning.

| | |
|---|---|
| Adam | Man |
| Seth | Appointed |
| Enos | Mortal |
| Cainan | Habitation |
| Mahalaleel | The Blessed God |
| Jared | Descend |
| Enoch | Teaching |
| Methuselah | His Death Shall Bring |
| Lamech | Captive |
| Noah | Comfort |

Putting the names of the patriarchs together we get a sentence that is prophetic of what God intends to accomplish in the human race by way of redemption. "Man appointed a mortal habitation (but) the Blessed God shall come down teaching (that) His death shall bring the captives comfort."

## Appendix 14: Nine Subdivisions of Genesis

The term "Generations" means simply beginning. The Book of Genesis is subdivided by the term that denotes the beginning of a new stage in the development of God's purposes for the human race. There are nine such subdivisions. These subdivisions are as follows:

1. The Generations of the heavens and the earth (Gen 1:1 - 2:4)

2. The Generations of Adam (Gen. 2:4 - 5:1)

3. The Generations of Noah (Gen. 5:1 - 6:9)

4. The Generations of the Sons of Noah (Gen. 6:9- 10:1)

5. The Generations of Shem (Gen. 10:1 - 11:10)

6. The Generations of Terah (Gen. 11:10 - 11:27)

7. The Generations of Isaac (Gen. 11:27 - 25:19)

8. The Generations of Jacob (Gen. 25:19 - 37:2)

9. The Generations of the Sons of Jacob (Gen 37:2 - Exd. 1:1)

# Appendix 15: The Bible is the Real School of Study for the Believer

Geology

Sedimentary Geology is explained by the Flood of Genesis. Glacial Geology is explained by the events that divided the world into continents in the days of Peleg.

Biology

Reproduction of living organisms is always "After its kind." This leaves no room for any evolution whereby one "kind" evolved into another. The genetic code that God put into the cells of every living creature has fixed the propagation of that creature so that it can reproduce only another of its kind.

Sociology

Man is designed by God to function together in harmony for the mutual benefit of all. God introduced into the race a series of divine institutions of marriage, family, nationalism, and human government for the orderly functioning of a healthy society.

Anthropology

Real understanding of why man is as he is and what makes him "Tic" is found only in the Bible. The existence and origin of the sin nature and the commands of God for harmonious functioning together gives man the resources by which he lives and can prosper on earth.

Political Science

God has a purpose for the nations but man has a different purpose. Politics can only be understood when one understands God's purpose.

Linguistics

Genesis gives us the origin and purpose for the different languages.

Psychology

The Bible gives us the information on why the makeup of the soul and the personality of man are as we find it.

Theology

Genesis gives us the introduction to God that we need in order to relate to Him as a loving creator and redeemer.

**"Let God be true, but every man a liar..." Rom 3:4**

## Appendix 16: The Curse Endured by both Man and the Savior

"<sup>17</sup> And unto Adam he said, Because thou hast hearkened unto the voice of thy wife, and hast eaten of the tree, of which I commanded thee, saying, Thou shalt not eat of it: cursed *is* the ground for thy sake; in sorrow shalt thou eat *of* it all the days of thy life; <sup>18</sup> Thorns also and thistles shall it bring forth to thee; and thou shalt eat the herb of the field; <sup>19</sup> In the sweat of thy face shalt thou eat bread, till thou return unto the ground; for out of it wast thou taken: for dust thou *art*, and unto dust shalt thou return."

**Genesis 3:17-19 (KJV)**

We note that the ground from which Adam was taken was cursed. We note also that it was cursed "…for thy sake." That is to say that God cursed the ground for Adam's good. Because of Adam's fallen state, Adam could no longer enjoy the good life of living in paradise. However, Adam was not alone in the enduring of that cursed ground. The Savior Himself also endured the curse. It is as if Jesus Christ the Creator / the Son of God said to the Father: "I have created them but they have fallen into sin and have become a fallen creature. Therefore I will enter into humanity and endure the curse with them so that I can redeem them back to us – the Godhead."

We see the Lord suffering in sympathy with His groaning creation in Romans Chapter 8 verses 16 through 25: "<sup>16</sup> The Spirit itself beareth witness with our spirit, that we are the children of God: <sup>17</sup> And if children, then heirs; heirs of God, and joint-heirs with Christ; if so be that we suffer with *him*, that we may be also glorified together. <sup>18</sup> For I reckon that the sufferings of this present time *are* not worthy *to be compared* with the glory which shall be revealed in us. <sup>19</sup> For the earnest expectation of the creature waiteth for the manifestation of the sons of God. <sup>20</sup> For the creature was made subject to vanity, not willingly, but by reason of him who hath subjected *the same* in hope, <sup>21</sup> Because the creature itself also shall be delivered from the bondage of corruption into the glorious liberty of the children of God. <sup>22</sup> For we know that the whole creation groaneth and travaileth in pain together until now. <sup>23</sup> And not only *they*, but ourselves also, which have the firstfruits of the Spirit, even we ourselves groan within ourselves, waiting for the adoption, *to wit*, the redemption of our body. <sup>24</sup> For we are saved by hope: but hope that is seen is not hope: for what a man seeth, why doth he yet hope for? <sup>25</sup> But if we hope for that we see not, *then* do we with patience wait for *it*."

| Sentence on Adam | Endured by Christ |
|---|---|
| The ground cursed | He became a curse for us (Gal 3:13). |
| Sorrow of Conscience | He became a Man of sorrows (Isa. 53:3). |
| Thorns and thistles | He wore a crown of thorns (Mt. 27:29). |
| Sweat on his face | He sweat great drops of blood (Luke 22:41) |
| The earth would reclaim his body. | He was three days and nights in heart of earth. |

## Appendix 17: The Firmament

"⁶ And God said, Let there be a firmament in the midst of the waters, and let it divide the waters from the waters. ⁷ And God made the firmament, and divided the waters which *were* under the firmament from the waters which *were* above the firmament: and it was so. ⁸ And God called the firmament Heaven. And the evening and the morning were the second day.

(Genesis 1:6-8)

The term "firmament" means simply "Open Space" God made an open space in which the sun, the moon, and the stars would exist.

The waters which were under the firmament would remain on the earth (Psalm 136:5-9).

The waters which were above the firmament would be moved out beyond the stellar universe to form the waters that divided the second heaven from the third (Psalm 148:4, 7 & 8).

The firmament "Heaven" is a reference to that which was above the earth's surface (Genesis 1:8 cf. Job 38:30; 37:18; Rev. 4:6; Job 26:7).

The term "the deep" is used of the water that covered the earth in Genesis 1:2 but it also refers to the water that comprises the water above the firmament. "Thou coverest it with the deep" (Psalm 104:5ff). The deep is still there but now there is a firmament in the midst of it.

The sun, moon and stars are in the open firmament (Genesis 1:14).

## Appendix 18: The Law of Cause and Effect

There is probably no greater example of the irrational approach to the various theories of origins than the failure to apply the law of cause and effect. People who are regarded as scientists talk about the "Big Bang" theory and picture the mass of the universe being originally at one point and then expanded to its present configuration. What is missing in this analysis is the fact that it does not account for the fact that the matter and energy had to have been brought into existence by a first cause that was bigger than itself.

The Law of Cause and Effect is presented here for the reader's consideration: Every phenomenon is an effect due to a cause. No effect is ever quantitatively greater than or qualitatively superior to its cause.

The First Cause of:                                   Must be:

    Limitless Space _____Infinite
    Endless Time_____Eternal
    Boundless Energy_____Omnipotent
    Universal Interrelationships_____Omnipresent
    Infinite Complexity_____Omniscience
    Moral Values_____Morally Perfect
    Human Responsibility_____Volitional
    Human Integrity_____Truthful
    Human Love_____Loving
    Life_____Living

# Appendix 19: The Scientific Method for Verification of a Theory

The Scientific Method for the verification of a theory requires four things. They are:

1.  Observation – You have to be able to collect empirical data to test the hypothesis.
    But neither creation nor evolution has ever been observed by any human being.

2.  Experimentation – You have to be able to study the theory in the laboratory.
    But the phenomena of the Universe's existence exceed any researchers lifetime.

3.  Reproduction of Results – You have to be able to reproduce the results.
    But neither creation nor evolution can be reproduced in the laboratory.

4.  Falsification – The theory must be able to be proven false (if it can be) in order for it to be proven to be true.
    But creation can not be refuted. Thus it is outside the realm of empirical science.

Thus it is "By faith that we understand that the worlds were formed by the word of God…" (Heb. 11:3)

## Appendix 20: The Source of the Water

In Genesis Chapter 7 we read "[17]And the flood was forty days upon the earth; and the waters increased, and bare up the ark, and it was lift up above the earth. [18] And the waters prevailed, and were increased greatly upon the earth; and the ark went upon the face of the waters. [19] And the waters prevailed exceedingly upon the earth; and all the high hills, that *were* under the whole heaven, were covered. [20] Fifteen cubits upward did the waters prevail; and the mountains were covered."

(Genesis 7:17-20)

We ask: Where Did the Water Come From?

- It did not rain before the flood (Gen. 2:5).
- The Rain did not come from the earth's atmosphere. The atmosphere as it exists today could not hold enough water to cover the highest mountain by a depth of about 25' (15 cubits in Genesis 7:20).
- In Genesis 7:11 we see that the water came from two sources:
    1. "The fountains of the great deep were broken up."
    2. "The windows of heaven were opened."
- The fountains of the great deep are apparently the points from which the subterranean water went up as a mist to water the earth (Gen. 2:5). These no longer existed after the flood because they were "broken up" at the time of the flood.
- The windows of heaven could be a reference to the collapse of the vapor canopy or more likely it refers to the water that is "above the firmament" (Gen. 1:7 & 8). Remember that the "firmament" of Genesis 1:9 & 10 provided and open space that encompasses the sun, moon and the stars (Gen. 1:14). The windows of heaven would be a reference to pathways through the stellar universe whereby some of the water that was above the firmament came back to the earth. This was some of the water that had inundated the earth in Genesis 1:2.

Where did the water go?
- In Genesis 8:1 we find that "God made a wind to pass over the earth and the water assuaged…"
- In Genesis 8:3, we see the results: "And the waters returned from <u>off</u> the earth continually…"
- God apparently returned the waters that came through the "windows of heaven" to where is was before the flood. Some of the waters that came from the fountains of the great deep are now in our oceans.

# Appendix 21: The Threefold Temptation of Eve

## The Temptation of Eve and the fall of Adam

Eve was tempted in a threefold manner. The first epistle of John makes mention of the threefold temptation that affects all of us saying "For all that is in the world, the lust of the flesh, the lust of the eyes, and the pride of life is not of the Father but is of the world." (1John 2: 16):

| **1John 2:16** | **(Gen.2:6)** |
|---|---|
| The Lusts of the Flesh | "…it was good for food…" |
| The Lusts of the Eyes | "…it was pleasant to the eye…" |
| The Pride of Life | "…a tree to be desired to make one wise…" |

Eve was deceived (1Tim. 2:12-15) but Adam flat out sinned (Rom. 5:12).
He could have taken it to God (Num. 30:6-16) but his problem was unbelief of the goodness of God.

---

## The Temptation of Christ and His Victory

Tempted in all points like as we (Heb. 4:14-15).

| **1John 2:16** | **(Matt. 4:3-8)** |
|---|---|
| The Lusts of the Flesh | "…stones be made bread…" |
| The Lusts of the Eyes | "…the kingdoms of the world and the glory of them…" |
| The Pride of Life | "…cast thy self down…" |

Our Lord's victory came in the words "It is written…" His victory came from His knowledge of the Word of God.

**Appendix 22:  The Bible Doctrine of Sin**

## When Does Temptation Become Sin?

Falling into sin involves a fivefold process of yielding to temptation:

Presentation:    The opportunity is set before you (Matt. 18:7).

Illumination:    The opportunity is pointed out to you as something attractive.

Debate:    You entertain the thought of doing it. (Sin starts here)

Decision:    You decide to go through with it. (It becomes sin here)

Action:    You actually do it.

### Appendix 23: Why Man Needs a Redeemer

1.        Sin had entered the human race
           Sin had to be removed.
           But man could not remove it.
           It was by man that sin entered.

2.        The Serpent had taken the kingdom that was to be given to man.
           It had to be won back.
           Man could not do that.
           Man had become the serpent's slave.

3.        Death had entered the human race.
           Death had to be defeated.
           Man could not do that.
           Man was now spiritually dead.

The term "seed" everywhere else in Scripture is used in reference to a man's seed. But in Genesis 3:16, it is used in reference to the seed of the woman. It is important to remember here that Eve was taken out of Adam before the fall. There is coming an offspring of the woman who would not have a human father. That one would be born without a sin nature. The sin nature is thus understood to be passed on through the male line. Thus Christ was born of a woman (Gal. 4:4) who was a virgin (Isa. 7:14) and was "the only begotten Son of God" (John 3:18). Thus He is the only person who is both God and man in one person. As such, He is the only avenue of approach for any man today to come to God; and then only through faith in the fact that He willingly surrendered His perfect life of perfect obedience to atone for the sins of His fellow men. What a wonderful Savior we have in our Lord Jesus Christ.

To understand the significance of Abraham in the Bible and in his relationship to all people of faith, we need to study the third chapter of Galatians. This appendix presents a study of that key chapter of scripture to give you the reader a grasp of his significance as the father of all who are of faith.

This is an annotated study of Galatians Chapter 3. Note that the letters designated in superscript in the Bible Text above (Even as Abraham[A]) corresponds to the annotation designated by the corresponding lettered note as for example ([A])

**Galatians 3:1-29 (KJV)** [1] O foolish Galatians, who hath bewitched you, that ye should not obey the truth, before whose eyes Jesus Christ hath been evidently set forth, crucified among you? [2] This only would I learn of you, Received ye the Spirit by the works of the law, or by the hearing of faith? [3] Are ye so foolish? having begun in the Spirit, are ye now made perfect by the flesh? [4] Have ye suffered so many things in vain? if *it be* yet in vain. [5] He therefore that ministereth to you the Spirit, and worketh miracles among you, *doeth he it* by the works of the law, or by the hearing of faith?

[6] Even as Abraham[A] believed God, and it was accounted to him for righteousness. [7] Know ye therefore that they which are of faith, the same are the children of Abraham. [8] And the scripture, foreseeing that God would justify the heathen through faith, preached before the gospel[B] unto Abraham, *saying*, In thee shall all nations be blessed. [9] So then they which be of faith are blessed with faithful Abraham.

[10] For as many as are of the works of the law are under the curse: for it is written, Cursed *is* every one that continueth not in all things which are written in the book of the law to do them. [11] But that no man is justified by the law in the sight of God, *it is* evident: for, The just shall live by faith. [12] And the law is not of faith: but, The man that doeth them shall live in them. [13] Christ hath redeemed us from the curse of the law, being made a curse for us: for it is written, Cursed *is* every one that hangeth on a tree:

[14] That the blessing[C] of Abraham might come on the Gentiles through Jesus Christ; that we might receive the promise of the Spirit through faith. [15] Brethren, I speak after the manner of men; Though *it be* but a man's covenant, yet *if it be* confirmed, no man disannulleth, or addeth thereto.

[16] Now to Abraham and his seed were the promises made. He saith not, And to seeds, as of many; but as of one, And to thy seed, which is Christ[E]. [17] And this I say, *that* the covenant, that was confirmed before of God in Christ, the law, which was four hundred and thirty years after, cannot disannul, that it should make the promise of none effect.

[18] For if the inheritance[F] *be* of the law, *it is* no more of promise: but God gave *it* to Abraham by promise. [19] Wherefore then *serveth* the law[G]? It was added because of transgressions, till the seed should come to whom the promise was made; *and it was* ordained by angels in the hand of a mediator. [20] Now a mediator is not *a mediator* of one, but God is one. [21] *Is* the law then against the promises of God? God forbid: for if there had been a law given which could have given life, verily righteousness should have been by the law. [22] But the scripture hath concluded all under sin, that the promise by faith of Jesus Christ might be given to them that believe.

[23] But before faith came, we were kept under the law[H], shut up unto the faith which should afterwards be revealed. [24] Wherefore the law was our schoolmaster *to bring us* unto Christ, that we might be justified by faith. [25] But after that faith is come, we are no longer under a schoolmaster. [26] For ye are all the children of God by

faith in Christ Jesus. [27] For as many of you as have been baptized into Christ have put on Christ. [28] There is neither Jew nor Greek, there is neither bond nor free, there is neither male nor female: for ye are all one in Christ Jesus. [29] And if ye *be* Christ's, then are ye Abraham's seed, and heirs according to the promise.

## Annotations on Galatians Chapter Three

[A]. Galatians is written by the apostle Paul to defend the Gospel of the Grace of God from the legalistic men from Judaea who would add works to the gospel. The particular concern was the issue of circumcision (Acts 15:1). In defending the Gospel of the Grace of God, Paul uses Abraham as an illustration of one who was justified by faith apart from works (Rom 4:1-4). We should not say that Abraham is the example of being justified by grace through faith apart from works because James used the same man Abraham as an illustration of how a man is justified by works and not by faith only (James 2:24). Remember that justification = being reckoned righteous = being fit to enter eternal life. To catch the point that the apostle is making here, we must remember that what the Gentiles have today in the Dispensation of the Grace of God is in no way inferior to what Israel had. To make that point, Paul points out that Abraham, the father of the Jewish nation, was justified by grace through faith apart from works and apart from the law while he was still a Gentile (i.e. before he received the rite of circumcision). The Gentile Abram was justified by faith without works in Genesis 15:1-6. The circumcised Abraham was justified by faith that was demonstrated by works in Genesis 22. The issue in Galatians is not how Israel in the Old Testament was really justified but how Gentiles are justified today. Paul is simply making the point that, because God knew that He would one day justify the Gentiles by faith apart from works in this present dispensation, He justified the Gentile Abram that way before he became the circumcised Abraham - the progenitor of the Hebrew nation. Therefore, what the Gentiles have today by grace through faith apart from works is superior to what Israel had under law. The Gentiles are thus instructed to stand fast in the liberty that they have in Christ. Also, the Circumcision had nothing to add to the Gentile who is justified by faith. We might ask "Why was Abraham justified by God by faith before He received the rite of circumcision?" The answer is that God wants Israel and the world to know that justifying men by grace though faith apart from human merit is His first choice and His preference.

[B]. In verse 8 we read that the Scripture "preached the gospel unto Abraham" that "…the blessing of Abraham might come to the Gentiles." What gospel did God preach to Abraham? If we go to the reference in Genesis 15:1-6, the gospel (good news) was simply this: *"This shall not be thine heir; but he that shall come forth out of thine own bowels shall be thine heir. 5: And he brought him forth abroad, and said, Look now toward heaven, and tell the stars, if thou be able to number them: and he said unto him, So shall thy seed be."* And the result *was "And he believed in the LORD; and he counted it to him for righteousness."* Abram simply believed a promise from God and God reckoned him righteous for the simple fact that he believed God. Though the content of the good news that the Scripture preaches to us is different than that preached to Abram, we none the less are justified the same way as Abram – simply by believing the good news. The good news preached to us is that "…Christ hath redeemed us from the curse of the law, being made a curse for us…" (vs 13).

[C] Verse 14 speaks of "the blessing of Abraham" that "…might come to the Gentiles." What is that blessing? It is obviously not the land of Canaan. That was given to Abraham's physical seed. The blessing of Abraham that we Gentiles share with Abraham is eternal life received on the basis of faith alone. God gave the land of Canaan to Abraham and his seed for an everlasting possession (Gen 17:8) but then tells him that he will die before he possesses it (Gen15:13-15). He would understand from this that there will be a resurrection before he inherits it. Today eternal life comes to the Gentiles through Jesus Christ the same way that it came to Abraham – by faith apart from works. Righteousness is imputed to us today the same way it was imputed to the Gentile Abraham.

> *Romans.4:16: Therefore it is of faith, that it might be by grace; to the end the promise might be sure to all the seed; not to that only which is of the law, but to that also which is of the faith of Abraham; who is the father of us all, 17: (As*

*it is written, I have made thee a father of many nations,) before him whom he believed, even God, who quickeneth the dead, and calleth those things which be not as though they were. 18: Who against hope believed in hope, that he might become the father of many nations, according to that which was spoken, So shall thy seed be. 19: And being not weak in faith, he considered not his own body now dead, when he was about an hundred years old, neither yet the deadness of Sara's womb: 20: He staggered not at the promise of God through unbelief; but was strong in faith, giving glory to God; 21: And being fully persuaded that, what he had promised, he was able also to perform. 22: And therefore it was imputed to him for righteousness. 23: Now it was not written for his sake alone, that it was imputed to him; 24: But for us also, to whom it shall be imputed, if we believe on him that raised up Jesus our Lord from the dead; 25: Who was delivered for our offences, and was raised again for our justification.*

**[D]**. What is the "...Promise of the Spirit..." referred to in Galatians 3:14? The promise of the Spirit in verse 14 is not the promise made to Abraham. The promise to Abraham was the land. We today have the promise of the Spirit. However, the promise to Abraham and the promise to us is none the less basically the same thing – i.e. eternal life. Abraham had the promise of eternal life to inherit the land. We have the promise of eternal life in the heavens (2Cor. 5:1). Our inheritance is eternal life in the heavens.

**[E]**. To whom was the promise made in verse 16? Paul says that the promise was not to all of the physical seed of Abraham but that the seed is only one – that seed being Christ. It is only to that of the physical seed that finds it's fulfillment in Christ that inherit the promise. Because the believing Gentiles today are baptized into Christ, they are also the seed of Abraham and heirs to eternal life by promise as Abraham was.

**[F]**. What is the inheritance? The inheritance (verse 18) is related to the promise. Just as the basic promise is eternal life while the specifics of the promise depends to who or of whom one is speaking, so too with the inheritance. The basic inheritance is eternal life (as with the question of the rich young ruler of Mark 10:17, 25). The specific inheritance for Israel is eternal life in the Kingdom of Heaven while the specific inheritance for members of the church which is Christ's Body is the Kingdom of God in the heavens.

**[G]**. Why was the law given? Verse 19 answers that question. There are two reasons – though we have to go elsewhere in Paul's epistles to understand them:

1. The law was added to the unconditional Abrahamic Covenant because of transgressions. Because of the nature of the natural man, the law had to be added to teach man of his need for Christ and the regeneration that comes from Him. Until the Holy Spirit is given through Christ, man (represented by Israel) remained a natural man. None of the works of the Holy Spirit regarding the regeneration of the individual believer happened to any man prior to the filling with the Holy Spirit at Pentecost and the eight different works of the Holy Spirit (e.g. Regeneration, circumcision, indwelling, sealing, baptizing, washing, etc.) presented in the Pauline epistles for the Dispensation of the Grace of God. Faith alone as the way of approach to God could not be revealed until Christ (the real seed) came. Prior to Pentecost, the Holy Spirit came upon certain men to do specific works (Numbers 11:17, 26; 27:18; Daniel 4:8; Ezekiel 2:2; 3:12; etc.) but the broad based, universal filling of the Holy Spirit did not happen until the offer of the kingdom to Israel in the Book of Acts.

2. The law was ordained by angels. That is, it was appointed by angels to keep Israel on track and to protect God's inheritance (which Israel was). Hebrews 2:2 speaks of the law as "...that which was spoken by angels." Colossians 2:18 speaks of the law as the worshipping of angels – i.e. the kind of worshipping that came from angels. The Lord was teaching angels something about his wisdom by imposing the system of worship that is called "the law." That's why "now unto the principalities and powers in heaven places might be known by the church the manifold wisdom of God..." (Eph 3:10). What angels and men are learning about the wisdom of God through the grace program of justification by faith apart from works is that grace can and will produce practical righteousness in man. Romans 8:4 tells us that grace can and will produce the righteous standard in man that the Law demanded of man but could not produce in man..

**[H]**. Was Israel really under law or did it just appear that they were and that they were justified by faith alone as Abraham was?

True righteousness has to be imputed to man because man can not produce true righteousness in his own strength. There was a righteousness of the law that God accepted (Deut. 6:25) while Israel was under law but that was only because of the blood sacrifice on the mercy seat (Ex. 30:6; Lev. 16:2, 14) to cover man's failure to keep it. However, the Old Testament sacrifices provided remittance for Israel's (and other Old Testament believers') sins only because God foresaw the real sacrifice: "...the righteousness of God which is by faith of Jesus Christ..." (Rom. 3:22) "whom God hath set forth to be a propitiation through faith in his blood, to declare his righteousness for the remission of sins that are past through the forbearance of God."(Rom. 3:25) An Israelite under the law was considered righteous as long as he walked according to the law as best they could. However, if he departed from his efforts to keep the law, he was no longer reckoned righteous but wicked. Hence, we find the conditional security of Ezekiel Chapter 18.

Galatians 3:23 "But before faith [i.e. faith as a means of justification apart from human merit] came we [Israel] were kept under law <u>shut up</u> onto the faith [the body of doctrine for today] which should afterward be revealed." Galatians 3:24 "But after that faith is come, we are not longer under a school master" [i.e. the law]. Galatians 3:11 says that the law is not of faith. Had Israel acted on faith, the nation would have objected to the conditions of the law and laid claim on the unconditional covenant that God made with Abraham. Israel was being tested by God when the law was offered to them. Israel failed the test and foolishly agreed (Exd. 19:8) to put themselves under a conditional covenant with God when they already had an unconditional covenant with Him through Abraham. See the Appendix 25 "Summary of the Covenants"

So did Israel have eternal security as we do today? No, Israel under the law had to maintain that conditional covenant until that law covenant was taken out of the way for them again so that only the unconditional covenant with Abraham remained. That will not happen until the day of Israel's atonement. The Book of Hebrews presents that opportunity to Israel. However, Israel under the Old Covenant was bound by that covenant and had to have the faith which that covenant demanded -- faith that would produce the works that it required. If he failed to walk after the statues of the law, it was because he did not have the faith in the conditional promises of the law and was no longer considered righteous until he again came in repentance and again embraced the law. We sure ought to appreciate grace when we consider Israel under law. Little wonder that Peter considered it a yoke that "neither we nor our fathers were able to bear." (Acts15:10).

## Appendix 25: The Covenants of God and the Mystery

What is a Covenant / a Testament?

**A Covenant:**   A contract between two parties.

             Two Types:     Conditional (Exd. 19:5)

                                "If you will…then I will"

Unconditional (Jer. 31:31 & 32)

                      "I WILL…"

**A Testament:**  A covenant that involves a death

                      All testaments are covenants but not all covenants are testaments

## Covenants that God has made:

1. Edenic      (Gen. 1:28)                    7 elements
2. Adamic     (Gen. 3:14)                    7 elements
3. Noahic     (Gen. 9:1; 11:10; 8:21)    7 elements
4. Abrahamic (Gen. 12:1;15:18)        7 elements
5. Mosaic     (Exd. 19:25;Mal. 5:17; Gal. 3:24)    3 parts
6. Palestinian (Lev. 26; Deut. 28:1; 30:3)    7 elements
7. Davidic    (2Sam. 7:16; 1Chr. 17:7; Psa. 89:27)    7 elements
8. New          (Heb. 8:8; Isa. 61:8; Rom. 11:25)    7 elements

## The Covenants of Promise Belong to Israel (Rom 9:4)

Table 9: **Israel and the Body of Christ**

| To Israel Belongs: | But We Today Have: (Rom. 15:27) |
|---|---|
| • The Adoption | • Adoption (Eph. 1:5) |
| • The Glory | • Glory (2Tim. 2:10) |
| • The Covenants | • The Mystery (Rom. 16:25) |
| • The Giving of the Law | • The Dispensation of Grace (Eph. 3:1) |
| • The Service of God | • Service (Rom. 11:31 & 32) |
| • The Promises | • Promises (Eph. 1:15) |

## The Covenants of Promise Include:

- Abrahamic
- Palestinian
- Davidic
- New Covenant

These were all made with Israel. But we have promises from God that He promised to Himself "… before the foundation of the world…" (Eph. 1:4; 2Tim. 1:9; Tit.. 1:2)

# 1. Edenic Covenant

**Seven Elements**
1. Replenish the Earth with a new order – Man (Jer. 4:25; Gen. 1:2)
2. Subdue the earth for human habitation (Heb. 2:6&7)
3. Have dominion over the earth (Matt. 25:34).
4. Eat Herbs and Fruit.
5. Till and Keep the Ground
6. Abstain from eating the Tree of the Knowledge of Good and Evil.
7. The Penalty of Violation is Death

# 2. Adamic Covenant

**Seven Elements**
1. The serpent cursed
   John 3:14 The Serpent in the Wilderness represents the curse that Christ bore (cf Num. 21:5-9).
   In 2Corinthians 5:21 we see Christ made to be sin for us.
   Brass a type of judgment / serpent a type of sin
   The serpent represents Sin being judged
2. The Proto Evangel (Gen. 3:15)
   The Seed Line Abel –Seth- Noah- Shem -Judah -David -Christ
3. Changed state of the woman (Gen. 3:16)
   Multiplied Conception
   Motherhood linked with sorrow
   Headship of the Man
   Entrance of sin requires headship
   Headship vested in the man (1Tim. 2:11-14; Eph. 5:22-25; 1Cor. 1:7-9)
4. The Ground Cursed – For Man's Sake (Gen. 3:17)
5. The inevitable sorrow of life. (verse 17)
6. The light work of Eden (Gen 2:15) changed to heavy labor (verses 18 19).
7. Physical death (verse14 cf Rom. 5:12-21)
   Spiritual death (Eph. 2:5 cf Gen. 2:17)

# 3. The Noahic Covenant

**Seven Elements in the Noahic Covenant**
1. Changed relationship of man and the earth (Gen. 9:2-4)
   The original commission of the Edenic Covenant remains (Gen. 9:1 cf 1:28-31).
   Dominion over the animal creation made more difficult (Gen. 2:2-4).
2. Seasonal order to life established
   Four Seasons:
   Of the Year
   Of life (Eccl. 3:1-15)
3. Human Government Established (Gen. 9:5&6; Acts 17:28; Rom. 13:1-6)
   Render to Caesar (Matt. 22:31).
   Be subject to the higher powers (Rom. 13:1-6).
   Obey magistrates (Tit. 3:1).
   These kingdoms will one day belong to Christ (Rev. 11:5; 12:10).
4. No future universal judgment by water (Gen. 8:21;9:11).

There was one previously (Gen. 1:2; Jer. 4:23).

There is a future destruction by fire (2Pet. 3:10).

5. Serving nature of Ham's posterity (Gen. 9:24,25).
6. Shem's posterity to be a blessing to God (Gen. 9:26,27).
7. Japheth's posterity to be enlarged (Gen. 9:27).

## The Institutions of God and the Covenants

The Edenic Covenant brought in three institutions

Main Purpose: A commission to be fruitful and multiply and replenish the earth.

**Institutions:**  1. Volition
2. Marriage
3. Family

**The Adamic Covenant brought another institution**

Main Purpose: Curse on creation for man's sake and promise of deliverance.

**Institution:**  4. Blood sacrifice

**The Noahic Covenant briught another institution**

Main Purpose: a means of dealing with the evil nature of man's heart

**Institution:**  5. Human Government and Cycles of Life

# 4. The Abrahamic Covenant

**Seven Elements:**

1. "I will make of thee a great nation …"
   - A natural posterity – "…As the dust of the earth"
     (Gen. 13:16; John 8:37)   i.e. the Hebrew People
   - A spiritual posterity – "…Look toward heaven…so shall thy seed be"
     (John 8:39; Rom. 9:7.8; Rom. 4:16; Gal. 3:6, 7, 29)
     …i.e. all men of faith
2. "I will bless thee…"
   - Temporal Blessings –
     Gen. 13:14, 15, 17; 15:18; 24:34)
   - Spiritual Blessings -
     (Gen. 15:6; John 8:56)
3. "I will make thy name great…"
4. "Thou shalt be a blessing"
   (Zech. 8:23)
5. I will bless them that bless thee…"
6. And curse them that curseth thee…"
   Deut. 30:7; Isa. 14:1, 2; Joel 3:1-9; Mic. 5:7-9;
   Hag. 2:22; Zech. 14:1-3; Matt. 24: 40, 45)
7. In thee shall all families of the earth be blessed.
   Genesis 3:15        Christ (Gal. 3:16; John 8:56-58)

**Three main parts to this Covenant:**

| Promises | [A Land] | [A Seed / A Nation] | [A Blessing] |
|---|---|---|---|
| Confirmed in | [Palestinian Cov.] | [Davidic Cov.] | [New Cov.] |

A note on the Abrahamic Covenant and us:

The apostle Paul (the apostle of the Gentiles, our apostle, the divinely appointed revealer of the mystery program and the dispensation of the grace of God) tells us that believers today are:

- Children of Abraham (Gal. 3:7).
- Abraham's Seed (Gal. 3:29).
- But never that we are children of the covenant as Israelites are (Acts 3:25).

Galatians 3:8 speaks of Scripture foreseeing someone. We ask:

1. What people were foreseen by the Scripture when the gospel was preached to Abram 24 years before he was circumcised and 430 years before the law was added to the gospel (Gal. 3:19)? It was the heathen (Gentiles) of Paul's day; you and me living in the dispensation of the grace of God.
2. What was foreseen? That the uncircumcised heathen would one day be justified by faith alone
   - without circumcision,
   - without the law,
   - without any religious observances;
   - just as Abram was, by faith without works.
3. We today are Abraham's seed not by being physically the seed of Abraham but by being baptized into Christ who is the physical seed of Abraham. This is done not according to the Covenant that God made with Abraham but according to the mystery revealed through the apostle Paul (Gal. 3:27-29).

# 5. The Mosaic Covenant

## Three Parts to the Mosaic Covenant
1. The Moral Law (Exd. 20:1-17)
2. The Civil Law (Exd. 21:1 – 24:11)
3. The Ceremonial Law (Exd. 24:12 – 31:18)

The Mosaic Covenant was:

- Conditional (Exd. 19:5).
- A Legally Binding Contract (Exd. 19:8).
- Unfriendly and Cold (Exd. 19:9-24).
- The Ministration of Condemnation (2Cor. 3:8; Rom. 4:15).
- Weak thru the Flesh (Rom. 8:3).
- Not of Faith (Gal. 3:12).
  It was not of faith because it was not based on the faith principle but on the performance principle. For Israel under Law it was faith with works (Mark 10: 17-21; Luke 10:25; James 2:22).
- Temporary (Gal. 3:19).
- A School Master (Gal. 3:24).
- A Taskmaster (Gal 3:26).
- Ordained by Angles in the hands of a mediator (Gal. 3:19; Heb. 2:2; Col. 2::18; Eph. 3:10).

The Mosaic Covenant and Us

The Law was taken out of the way for us (Col. 2:14) .

We are free from the Law (Gal. 5:1).

Grace consistently applied will produce in us that which the Law demanded but could not produce (Rom. 8:4).

The Law and Israel's Future

> The Law was supplemented by the Palestinian Covenant (Duet. 29:1; 30:1-5). The unconditional Palestinian Covenant will give to Israel what they failed to gain as a result of their failure under the law.

# 6. The Palestinian Covenant

The Covenant that we call the Palestinian Covenant is found in Deuteronomy 28:63 thru 30:9. It is so called because it gives the conditions under which Israel will enter the Promised Land in spite of the fact that they disqualified themselves from entering by their failure under the Mosaic Covenant. Israel has never yet possessed all of the land that was promised them in Genesis 15:18 and Numbers 34:1-12.

God through Moses made the Palestinian Covenant some 40 years after the Mosaic Covenant (the Law) was made. The Law was conditional with potential for both blessing and cursing.

Table 10: **The Blessing and Cursing of the Law**

| The Blessings if the Law kept (Deut. 28:3-6) | Cursing if the Law broken (Deut. 28: 16-19) |
|---|---|
| 1. Blessed in the City | 1. Cursed in the City |
| 2. Blessed in the Field | 2. Cursed in the field |
| 3. Blessed in the fruit of thy body | 3. Cursed in the fruit of thy body |
| of the ground | The fruit of thy land |
| of thy cattle | The fruit of thy kine |
| | The fruit of thy sheep |
| 4. Blessed shall be thy basket | 4. Cursed shall be thy basket |
| Thy store | Thy store |
| 5. Blessed shalt thou be when thou comest in | 5. Cursed shalt thou be when thou comest in |
| 6. Blessed shalt be when thou goest out | 6. Cursed shalt thou be when thou goest out |

Because God knew that Israel would fail to keep the Law, God laid the unconditional Palestinian Covenant "beside" the conditional covenant of the Law. The Palestinian Covenant allows God to give Israel the land in spite of the fact that thy would be disqualified by the Law Covenant

- Israel entered into a legal contract with the Law (Exd. 19:12).
- But the Law is not of faith (Gal. 3:10 & 1).
- Had Israel operated on faith, they would have relied on the unconditional covenant that God made with them through Abraham (Gen. 12:1-3).
- Israel's problem was the same one that we tend to have; pride in thinking that they could live a righteous life in the strength of their own flesh (Hos. 13:1).
- But God will one day come to Israel's rescue (Hos. 13:9).
- Because of the broken Law and the Palestinian Covenant, Israel would go from being God's people -- To being "not God's people" (Hos. 1:0) -- To being again "God's People" (Hos. 1:10; 2:23 cf Rom 9:25).
- Note Moses' intercession for Israel in Exodus 32:30-33. Moses starts to understand that Israel's hope is in the unconditional covenant that God made with Abraham and not the conditional covenant that God made with Israel through him at Mount Horeb.

**There are Seven Parts to the Palestinian Covenant:**
1. Israel will be dispersed for their failure under the Mosaic Covenant (Deut. 28:63-68; 30:9).
2. There will be a future repentance of Israel while in dispersion (Deut. 30:2).
3. The Lord will return to Israel (Deut. 30:3; cf Amos 9:9-14; Acts 15:14-17).
4. Israel will be restored to the land (Deut. 30:5; cf Isa. 11:1; Jer. 23:3-8; Ezk. 37:21-25).
5. The Nation will be converted to the Lord (Deut. 30:6 cf. Rom. 1:26, 27; Hos. 2:14-16).
6. Judgment will be brought upon Israel's oppressors (Deut. 30:7; Isa. 14:1,2; Joel 3:1-8; Mt 25:31-46).
7. There will be a national prosperity for Israel from that point on (Deut. 30:9; Amos 9:1-14).

# 7. The Davidic Covenant

The Covenant that God made with King David is found in 2Samuel 7:12-16. The Davidic Covenant sets forth the promises upon which the promised kingdom would be established. This covenant is (as are the Abrahamic, the Palestinian, and the New Covenants) an unconditional covenant.

**It involves a five-fold promise that Jehovah made to David:**
1. "I will set up thy seed after thee [i.e. a family] which shall proceed out of thy bowels." (vs. 12)
2. "...I will establish his kingdom." (vs. 12)
3. "He shall build an house for my name [fulfilled in Solomon and eventually again in Christ], and I will stablish the throne of his kingdom forever." (vs.13)
4. "I will be his father and he shall be my son. If he commits iniquity, I will chasten him with the rod of men, and with the strips of the children of men."
5. "But my mercy shall not depart away from him, as I took it from Saul, whom I put away before thee. And thine house and thy kingdom shall be established forever before thee: thy throne shall be established forever." (vs. 16)

Note that four things are involved:
    1) His Seed. 2) His House, 3) His Throne, and 4) A Kingdom
- David would die but his seed would always sit on the throne of Israel (vs 23). Therefore, Mathew, who presents Christ as the King, starts out by referring to Him as "the son of David, the son of Abraham..."
- This covenant is often referred to as "The sure mercies of David." ( Psalm 89:3, 4, 20-37; Jer. 33:19-26; Psalm 132:10-14)
- Israel's future rests in David's seed and the Davidic Covenant (Isa. 54:1-17).
- The throne of David would be given to Christ (Luke 1:32,33).
- Christ was raised from the dead to sit on David's throne (Acts 2:30).
- Today, while God is "visiting the Gentiles" (Acts 15:14), the throne of David does not exist. However, it will one day be reestablished (Acts 15:15 &16 cf Amos 9:11).
- The hierarchy of the Davidic Kingdom:
  1. Christ will be King of kings and Lord of lords (Rev. 17:14).
  2. David will reign over Israel under Christ (Jer. 23:6-8; 30:9; Ezek. 37:34).
  3. The Twelve Apostles will Judge Israel under David (Isa. 32:1; Matt. 19:28).
  4. The twelve Tribes will rule over the Gentile Nations (Deut. 32:8).
  The earth will then be full of righteousness (Isa. 11:4; 26:9; etc.).

# 8. The New Covenant

The New Covenant is promised in the Old Testament Scriptures in Jeremiah 31:31 and Ezekiel 36:27 and other passages. However, we find blessings of it listed Hebrews: 8-13. The Book of Hebrews is written primarily to present the New Covenant to Israel. The Gentiles were "…strangers from the covenants of promise…" (Eph. 2:12) and thus strangers from the New Covenant that God would make with Israel. However, We have a body of doctrine that was ordained "…before the world began…" that gives us special blessings from God. That body of doctrine is called the mystery and contains blessings given uniquely to us living in the dispensation of the grace of God.

**There are Seven Elements to the New Covenant**
The New Covenant that will be made with Israel and with Judah (Heb. 8:8-12) has seven parts to it:
1. God will put His law in their minds.
2. He will write them is their hearts.
3. He will be their God.
4. They will be His people.
5. They will not need teachers.
6. They will all have supernatural knowledge of God.
7. God will forgive their sins and have no more remembrance of their shortcomings.

Seven facts that we find in Hebrews about the New Covenant:
1. It is better than the Mosaic Covenant (Heb. 7:19).
2. It is based on better promises (Heb. 8:10, 12 cf Exod. 19:5).
3. It has a better motivation
    - The Old Covenant -- Was based on fear (Heb. 2:2; 12:25-27)
    - The New Covenant -- Is based on love and grace (Heb. 8:10).
4. It is a personal relationship rather than a legal contract (Heb. 8: 10).
5. It involves a complete removal of sin rather than a temporary covering (Heb. 8:2; 10:17; cf 10:30).
6. It involves an accomplished redemption (Matt. 26:27.28; Heb. 9:1, 12, 18-23).
7. It secures Israel's future (Jer. 31:31-46; Ezek. 36:27).

The New Testament and Us (Members of the Body of Christ)
Paul our apostle tells us that God "…hath made us able ministers of the new testament; not of the letter but of the spirit…" (2Cor. 3:6). The New Covenant is promised to Israel and will be made with Israel. Though we are not under the New Covenant, we have the spiritual benefits of it by God's grace. Let's examine them in light of the New Covenant with Israel:
1. Though God did not put His law in our minds, we can have the mind of Christ as we mature under grace (1Cor. 2:16).
2. Though God did not write His law in our hearts, we are constrained by the love of Christ (2Cor. 5:14).
3. God is our God today (2Cor. 6:16).
4. We are His people today (2Cor. 6:16).
5. Though we still need teachers today (1Tim. 5:17), we have the completed word of God whereby we can know God even as we are known (1Cor. 13:12).
6. Though we do not automatically know God, we each can "…all come in the unity of the faith, and of the knowledge of the Son of God, unto a perfect man, unto the measure of the stature of the fulness of Christ:" (Eph. 4:13).
7. Just as Israel's sins will one day be forgiven, so it is that God has imputed the righteousness of Christ to our spiritual bank account (2Cor. 5:21; Rom. 3:22).

The Mystery -- is not a Covenant that God promised to man but a work of Grace that God promised to Himself before the world began.

### The Mystery concerns the Dispensation of the Grace of God

The information that God has given to us to govern how we who live in the dispensation of grace are to relate to Him is contained in the Pauline Epistles. We today have a relationship to God that is not detailed in any of the covenants of promise that God made with Israel but rather is laid out for us in a body of doctrine called "The preaching of Jesus Christ according to the revelation of the mystery." (Romans 16:25) It is called the mystery because it is "…the hidden wisdom, which God ordained before the world unto our glory…" but was not revealed until the time was right to do so. The time for the revelation of this hidden wisdom came when Israel rejected the offer of the Kingdom in Acts Chapter 7, with the stoning of Stephan. The next event in prophecy after the rejection of the offer of the kingdom should have been the seventieth week of Daniel (or what we call the tribulation period). However, instead of the Lord returning to bring the seventieth week of Daniel Chapter 9, He came to save Saul of Tarsus and to reveal the mystery through him.

### There are seven elements to this body of doctrine called the mystery.

They are:

1. The Dispensation of Grace - The beginning of the dispensation of the grace of God in which there is no unpardonable sin. In Matthew 12:31-32, we see the Lord introduce the concept of the unpardonable sin – that being the blaspheming of the Holy Spirit. It could not be forgiven "neither in this world nor in the world to come." "This world…" referred to the gospel era while "the world to come…" referred to Pentecost and the Seventieth Week of Daniel Chapter 9 when the Kingdom was being (and will be again) offered to Israel in which God was proclaiming "the acceptable year of the Lord" (Isa. 61:2; Luke 4:19). Saul of Tarsus along with the other leaders of Israel committed that sin of blasphemy of the Holy Spirit. Saul of Tarsus could, however, be forgiven in spite of having committed the unpardonable sin of blaspheming the Holy Ghost (Matthew 12:31and 1Timothy 1:14) because God inserted another dispensation (the dispensation of the grace of God) between the two (between this world and the world to come) in which there would be no unpardonable sin. Paul tells us that it was through him that Jesus Christ revealed the mystery involving the dispensation of grace (Ephesians 3:3 & 4; Colossian 1: 26 & 27).

2. The Present Temporary Partial Blindness of Israel (Romans 11:25) -- Israel is today in partial blindness because they as a nation rejected the offer of the kingdom made in Acts. It is a partial blindness because there are some Israelites getting saved today but they are saved by coming to God as a Gentile would. It is also only temporary because it is a blindness that will last until the fullness of the Gentiles is come in (Romans 11:25). The fullness of the Gentiles is a reference to the completion of the calling out of the Body of Christ. When the church which is His body is complete, it will be raptured to heaven. The rapture (1Thessalonians 4:15-17; 1Corinthians 15:51) will close the dispensation of grace and will take the church the Body of Christ to its eternal home in the heavens (2 Corinthians 5:1).

3. The Gospel - The truth about what was actually accomplished on the cross of Calvary is one of the elements to the mystery revealed through Paul. It is what Paul called "the mystery of the gospel for which he was an ambassador in bonds…" (Ephesians 6:19 & 20). Paul refers to this as "My Gospel" (Romans 2:16; 16:25), the gospel of his Son (Romans 1:9); the gospel of Christ (Romans 1:16); the gospel of peace (Romans 10:15) and the gospel of God (Romans 15:16). This gospel of Calvary is clearly laid out in detail in the Bible first in Romans 3: 20 – 25.

4. The Rapture - The truth concerning the fact that there is now with the church which is Christ's body a new elect agency – the church which is Christ's Body (a Gentile church). It is through this new elect agency that God will reconcile the heavenly places to Himself. Involved in this will be a rapture which catches the members of the Body of Christ up to heaven where the individual members will have resurrected (or raptured) bodies that are eternal in the heavens (2 Corinthians 5:1). With this addition of this mystery to God's program for the ages, we see how it is "That in the dispensation of the fullness of times he might gather together in one all things in Christ, both which are in heaven and which are on earth…" (Ephesians 1:9 and 10 and Colossians 1:14).

5. The Great Mystery - Involves the forming of a unique elect agency that is similar to the marriage relationship (Ephesians 5:32) to form the one new man of Ephesians 2:15. This one new man is composed of Jews and Gentiles joined together in one body that is "…flesh of his flesh and bone of his bone…" This one new man is Christ according to the revelation of the mystery.

6. The Mystery of Godliness - This one new man comprises the elect agency through which God's character and personality is manifest in the world today. Godliness is manifest in the world today through the church which is Christ's body as it functions in local churches (1 Timothy 3:16).

7. The Mystery of Iniquity - While God's character and personality is manifest in the church, Satan's character is manifest in the world of the unsaved masses which carries out his plan in the earth (2 Thessalonians 2:7).

The apostle Paul has much to say about the Law in Galatians especially as it relates to us today in the dispensation of grace. Let's look at Galatians chapter 3 on the law:

Galatians 3:11-27 (KJV) [11] But that no man is justified by the law in the sight of God, *it is* evident: for, the just shall live by faith. [12] And the law is not of faith: but, The man that doeth them shall live in them. [13] Christ hath redeemed us from the curse of the law, being made a curse for us: for it is written Cursed *is* every one that hangeth on a tree: [14] That the blessing of Abraham might come on the Gentiles through Jesus Christ; that we might receive the promise of the Spirit through faith. (Galatians 3:11-14)

The term "The just shall live by faith" is found in four places in the Bible (Habakkuk 2:4; Romans 1:17; here in Galatians 3:11; and in Hebrews 10:38). It is a statement to the effect that it is faith that actually saved people whether under Law in the Old Testament, under grace today, or in the coming kingdom under the New Covenant. Under the Law, people performed the requirements of the Law but understood that it was their faith in what God said that actually saved and not the performance of the works of the Law. The redeeming work that Jesus accomplished on the cross removed the curse of the Law for us. Paul is pointing out here that the cross removed the curse of the conditional covenant of the Law that was added to the unconditional covenant that God made with Abraham 430 years earlier.

[15] Brethren, I speak after the manner of men; though *it be* but a man's covenant, yet if *it be* confirmed, no man disannulleth, or addeth thereto. (Galatians 3:15)

Even in contracts between men, a contractual agreement is unalterable once it is made and agreed upon by both parties.

[16] Now to Abraham and his seed were the promises made. He saith not, And to seeds, as of many; but as of one, And to thy seed, which is Christ. [17] And this I say, *that* the covenant, that was confirmed before of God in Christ, the law, which was four hundred and thirty years after, cannot disannul, that

it should make the promise of none effect. [18] For if the inheritance *be* of the law, *it is* no more of promise: but God gave *it* to Abraham by promise. (Galatians 3:16-18) (KJV)

God made an agreement (an unconditional covenant) with Abraham and Abraham's seed. Here the apostle tells us that it is not with the seeds (plural) but with Abraham's seed (singular). That seed is Christ. The blessing of that covenant will go to the multiplied seed of Abraham but only to those of his multiplied seed who will find the fulfillment of the promise in Christ. The covenant of the Law which came 430 years later cannot nullify the unconditional covenant made with Abraham. The inheritance in view here is eternal life.

[19] Wherefore then *serveth* the law? It was added because of transgressions, till the seed should come to whom the promise was made; *and it was* ordained by angels in the hand of a mediator. [20] Now a mediator is not *a mediator* of one, but God is one. [21] Is the law then against the promises of God? God forbid: for if there had been a law given which could have given life, verily righteousness should have been by the law. (Galatians 3:19-21)

Why then did God give the Law? He added it because of Israel's transgression – her sin. Israel had to learn that they were as much in need of redemption as were the Gentiles. Therefore, God gave them the Law as a means of conviction. Performance of the Law could not save them but it served to point them to the blood sacrifice that did cover sin until the real sacrifice that got the job of redemption done was made by Christ.

[22] But the scripture hath concluded all under sin, that the promise by faith of Jesus Christ might be given to them that believe. [23] But before faith came, we were kept under the law, shut up unto the faith which should afterwards be revealed. [24] Wherefore the law was our schoolmaster *to bring us* unto Christ, that we might be justified by faith. [25] But after that faith is come, we are no longer under a schoolmaster. [26] For ye are all the children of God by faith in Christ Jesus. [27] For as many of you as have been baptized into Christ have put on Christ. (Galatians 3:22-27)

The scripture (represented by the Law) concluded all under sin that the promise of eternal life might be available to all by the faithfulness of Christ to accomplish redemption by His work of the cross. From the time of the giving of the Law until the revelation of the mystery of the gospel given through Paul, the human race was under the Law. The Law was then a school master to instruct people of their need of a redeemer. The faithfulness of Jesus on the cross has now opened up the way for us to be children of God. When we trust Jesus Christ and His work of redemption, the Holy Spirit baptizes us into an eternal spiritual relationship with Him to make us children of God.

**In summary**

A study of the Covenants that God has made with men is important in understanding God's plan for man and particularly God's purpose for Israel. Israel is God's Covenant people. He dealt with that nation on the basis of covenants. The apostle Paul says of us Gentile members of the Body of Christ that we "…were without Christ, being aliens from the commonwealth of Israel, and strangers from the covenants of promise, having no hope, and without God in the world…" (Ephesians 2:12). Peter on the other hand tells Israel: "…Ye are the children of the prophets, and of the covenant which God made with our fathers…" (Acts 3:25). Paul again says in Romans 9:4 that the adoption, the glory, the covenants, and the giving of the law all pertain to Israel. We looked at what God calls the Mystery as a body of doctrine that governs how God interacts with men today in the dispensation of grace. The program called the Mystery pertains to us who live today in the dispensation of grace. However, we must first get a handle on the covenants that God made with men that pertain to God's earthly people.

There are some covenants that do apply to the nations at large. If you regard any agreement that God made with men as a covenant, then there are eight covenants that can be identified in Scripture. They are:

| | | |
|---|---|---|
| Edenic | (Genesis 1:28) | 7 elements |
| Adamic | (Genesis 3:14) | 7 elements |
| Noahic | (Genesis 9:1; 11:10; 8:21) | 7 elements |
| Abrahamic | (Genesis 12:1; 15:18) | 7 elements |
| Mosaic | (Exodus 19:25; Malachi 5:17; Galatians 3:24) | 3 parts |
| Palestinian | (Leviticus 26; Deuteronomy 28:1; 30:3) | 7 elements |
| Davidic | (2Samuel 7:16; 1Chronicals 17:7; Psalm 89:27) | 7 elements |
| New | (Hebrews 8:8; Isaiah 61:8; Rom 11:25) | 7 elements |
| The Mystery | (The Pauline Epistles) | 7 elements |

The covenants that pertain exclusively to Israel are the Abrahamic, the Mosaic, the Palestinian, the Davidic, and the New Covenants. Of these five covenants, four were unconditional while one (the Mosaic) was conditional. It was conditioned upon Israel living up to the terms of walking according to its conditions and precepts.

## The Mystery

The Body of Doctrine that defines God's program for this present dispensation of grace is in the Pauline epistles and is called the Preaching of Jesus Christ according to the revelation of the mystery. This body of doctrine is not called a covenant per se in the Bible. The mystery (like each of the covenants) also has seven elements to it. Each element is said to be a mystery in itself. The elements to the mystery are:

1. The mystery of the gospel (Ephesians. 6:19; 1Corinthians 2:7)
2. The Great Mystery of Christ and the Church (Ephesians 5:23)
3. The mystery of the present, temporary, partial blindness of Israel (Romans 11:25)
4. The mystery of the rapture (1Corrhians 15:51)
5. The mystery of godliness (1Timothy 3:16)
6. The mystery of iniquity (2Thessalonians 2:7)
7. The mystery of the ultimate reconciling of things back to God (Ephesians 1:9 and 10)

For a further study of the body of Bible Doctrine called the Mystery, the reader is encourage to go to the Book *"You and Your Creator"* also by the same author.

# Appendix 26: From History to Prophecy it is all His Story

Table 11 From History to Prophecy it is His Story

| For by him were all things created (Col. 1:16)<br>He is before all things, and by Him all things consist | All things subdued unto Him (I Cor. 15:27)<br>All things gathered together in Christ (Eph. 1:10) |
|---|---|

| History – God's record of the past | | Declaring the end from ancient times (Isaiah 46:10)<br>I have declared the former things from the beginning (Isaiah 48:3)<br>In the latter days ye shall consider it perfectly (Jer. 23:20)<br>I have showed thee new things from this time (Isaiah 48:6) | | Prophecy – God's story of the future |
|---|---|---|---|---|
| | The Creation of Heaven and Earth (Gen 1:1) | | The New Heaven and New Earth (Rev. 21:1) | |
| | The first rebellion – Satan and angels (Isa 14:28; Ezek. 28:14)<br>The first judgment – chaos (Gen. 1:2) | | The final rebellion – Satan and men<br>The final judgment – fire (Rev. 21:8) | |
| | The Earth made ready for man (Gen. 1:3-31) | | The Earth a perfect habitat for man (Rev. 22:1-7) | |
| | The first man and his Bride (Gen. 2:18-25) | | The last Man and His Bride (Rev. 21:9-21) | |
| | The subjection to Satan (Gen. 3:1-19) | | The subjecting of Satan (Rev. 20:10) | |
| | The Earliest Gospel (Gen. 3:15)<br>Universal rebellion (Gen. 6:1-7)<br>Judgment by water (Noah –Gen. 6:8-22)<br>The earth purged by water (Gen. 7:17-24)<br>Governments setup (Gen. 9:5-7) | | The Everlasting Gospel (Rev. 14:6)<br>Universal rebellion (Rev. 20:8)<br>Judgment by fire (2Pet. 3:7)<br>The floor purged (Matt. 3:12)<br>Kingdom setup Perfect Government | |
| | Institution of Babylon<br>Idolatry invented (Gen. 11:1-4)<br>Nations scattered (Gen. 11:5-9) | | Destruction of Babylon (Rev. 18:2)<br>Idolatry ended  Rev. 9:20; 21:8)<br>Nations gathered (Rev. 16:4; 20:8) | |
| | Call of Israel (Gen 12:1 thru Duet.)<br>Blessing on Israel (I and II Sam)<br>Declension in Israel (I & II Kings)<br>Judgment on Israel (Isaiah, Jer. Ezek.) | | Restoration of Israel (Rev. 5:10)<br>Judgment of tribulation<br>Repentance of the nation (Rev. 7:4)<br>Blessing of the nation (Rev. 21) | |
| | The times of the Gentiles begin (Dan, Ezra, Neh.) | | The times of the Gentiles end (Luke 21:24; Rev. 11:5) | |
| | The first advent of Christ to the manger | | The second advent of Christ to the throne | |
| | Ministry of Christ<br>    The Truth<br>      His rejection and death<br>      His resurrection and ascension | | Ministry of Anti-Christ<br>    The Lie<br>      His reception and reign<br>      His destruction and doom | |
| | The Spirit poured out (Acts 2:17)<br>Second coming in view<br>The fall of Israel | | The Spirit again poured out (Rev. 19:10; 22:17)<br>Second coming in view<br>Rise of Israel | |
| | The Mystery revealed (Eph. 3)<br>The Body called out (Eph. 2:11-18)<br>Gentiles brought in (Rom. 11:16-25) | | The Mystery ended with the rapture (2Thess. 2:7)<br>The Body caught up (2Thess. 4:15)<br>Gentiles cut off (Rom. 11:26) | |

## A Study in Genesis
## From Adam to Abraham

The Book of Genesis is the Book of the origin of the universe in general and most significantly of the origin this earth and of man's divinely appointed place in the earth. This study delves into Genesis in search of truth that will establish in the mind of the Bible believer the absolute authority of the Word of God as divine revelation of truth to man. This study is the fruit of the author's prayerful search of the Bible to find the information that will fix in the mind and heart of the serious student of the Bible the infallibility of the scripture as the sure anchor for the soul. It is the need of every human heart to be established in that faith that promises eternal life as a gift of God's grace. The lost man when presented with the prospect of receiving the greatest gift any man can have – the gift of eternal life, is compelled to find in the Bible that sure anchor for his soul. In the quest for such security of the soul, the spirit of man needs to know that the Word of God will pass every test that man can give it.

This study brings the principles of science and engineering into play in the quest to find answers to some of the perplexing questions on origins in Genesis. It seeks to answer the questions such as how the laws of science come into play in creation and how what we observe in the geologic features of the earth squares with information in Genesis. The author answers for the Bible student questions such as: Was the flood of Noah's day truly world wide? What really brought on the glaciers? What actually gave us the fossil fuel reserves that power society today? What were the forces that divided our earth into continents? This study will fortify the Bible believer with information that will enable him to resist the challenges that have been raised by those who seek to undermine the faith of Bible believing Christians. Having heard the gospel that this great gift of eternal life is extended to all men, the Bible student needs to satisfy himself in his mind that the Bible is true and accurate in every detail from beginning to end. This study will do that for him

### ABOUT THE AUTHOR

Michael J. Tiry came to know the Lord Jesus Christ as his personal Savior at the age of twenty nine while in the midst of a successful career as an engineer. Michael served in the United States Civil Service as a professional engineer for 25 years. After 25 years with the civil service, Michael also started, owned and operated a private engineering company. While engaged in a career as an engineer, he also was involved with other men in the founding a local Bible believing church. His deep appreciation for having the assurance of eternal life, his passion for study, and his quest for truth compelled him to search deeply into the Bible with a desire to learn its truth that he might present the riches of God's grace to others. Over the last forty five plus years Michael has been involved in itinerate preaching, a church planting ministry, and a teaching and preaching ministry at Berean Bible Church in Chippewa Falls, Wisconsin. Michael also serves Berean Bible Church as director of the Timothy Institute – a Bible curriculum designed to prepare men for leadership in local churches. Additionally, Mike has been active over a span of twenty three years in a prison ministry. Michael and his wife of forty five years (Linda) have raised five daughters.

### OTHER BOOKS BY THE SAME AUTHOR

Michael has written over sixteen books which are used as study guides in the Timothy Institute. This book is one of the four that has been published. Others include "You and Your Creator" and "More than Conquerors" ("Super Abounding Grace) and "A Study in the Revelation." All are available through Barnes and Noble, Amazon and other book distributors.